The Ultimate Guide to STARGAZING

Second Edition 2023

VOLUME I

MAXIMIZING YOUR ENJOYMENT OF
STARGAZING

The Ultimate Guide to Stargazing - Volume 1

Second Edition 2023

ISBN: 978-0-6454669-1-1 (paperback)

Copyright © Dreamtech Designs & Productions Pty Ltd & Gregg Thompson 2022. All rights are reserved. The moral right of Gregg Thompson to be identified as the author of this work has been asserted.

The Ultimate Guide to Stargazing contains content material which is owned by or licensed to and are the copyright of Dreamtech Designs & Productions Pty Ltd.

The content material includes, but is not limited to, text, diagrams, design, layout, look, appearance, logos, trademarks and graphics. Without limiting the rights under copyright laws, you are not permitted to reproduce the content material, documents or information in the Ultimate Guide to Stargazing for any purpose or for use by any third party without the express written consent of Dreamtech Designs & Productions Pty Ltd. In particular you are not permitted to reproduce, republish, upload, download, transmit or store electronically or otherwise or distribute any part of The Ultimate Guide to Stargazing, whether for commercial exploitation or otherwise, other than as specifically licensed by us upon purchase of the Ultimate Guide to Stargazing.

Purchase of the Ultimate Guide to Stargazing is for use by the purchaser only.

Limited consent is hereby granted to enable content from the Ultimate Guide to Stargazing to be used for quotation of brief passages in reviews. We reserve the right to revoke any consent as we see fit. Please contact us to request any further consents.

We expressly reserve the right to take action against you if you breach any of these terms. You may be liable to criminal prosecution and civil claims for damages.

Although the author and publisher have taken all reasonable care in preparing this book, no warranty is given about the accuracy or completeness of its content, nor for any content linked or referred to in it, nor for any postings, message boards, Internet links, online commentary or the like. No responsibility is accepted and all liability arising from its use is disclaimed to the maximum extent permitted by law.

The author and the publisher have, to the best of their knowledge, refrained from using any copyrighted material unlawfully or without consent and have properly attributed such rights where appropriate. If you find any unindicated material protected by copyright, the copyright could not be determined. In the case of any unintentional copyright violation the author will remedy that by removing the material or properly attributing it.
Kindly contact us on: info@theultimateguidetostargazing.com

First published by Dreamtech Designs & Productions Pty Ltd 2018 ©

DEDICATION

I dedicate this book to my family. Firstly, to my wife Deborah for her loyal support and for the encouragement she has given me over the many years it has taken to do the research, produce the imagery and write *STARGAZING's* volumes. I have greatly appreciated her advice and discussions on subject matter and content as well as the time she invested in assisting with editing. I am also in debt to my very supportive son Tane for publishing and marketing this book, as well as his feedback on certain subjects and his help with editing over the years. If it was not for his dedication and his expertise in managing our tourist attractions, I would not have had the freedom or the time to do the massive amount of research required, or the time to write this book.

I must give a special thanks to my father and mother who both did so much to encourage my endeavors in astronomy and in the other sciences from when I was only eight. My father bought me expensive books on astronomy and the other sciences, as well as buying for me my first telescopes, a microscope, a chemistry set, electricity and magnetism experiments, gliders, model aeroplanes, and model boats, as well as a large trains set, and many puzzles – all of which I treasured. In my youth before I was old enough to drive, my wonderfully considerate mother would drive me and my telescope out to my uncle's farm in the country, so I could observe deep sky objects in their full glory.

Finally, I wish to thank my beautiful and intelligent granddaughters Aisja and Sienna, who have given me much inspiration and joy and who are the future of our family. They will live in the world of increasingly exciting inventions and discoveries, incredible concepts and extraordinary change, as explained in this book.

The Milky Way rising behind a quaint stone church on a hilltop in New Zealand photographed by Nico Babot.

CONTENTS

ACKNOWLEDGMENTS ... I
ABOUT THE AUTHOR ... III
AN INTRODUCTION TO THE STARGAZING VOLUMES ... XIII

CHAPTER 1: THE WONDER AND JOY OF STARGAZING ... 1
   WHY IS STARGAZING SO EXCITING TO SO MANY PEOPLE? ... 2
   STARGAZING'S 'WOW!' FACTOR ... 3
      Ten-Thousand Star Accommodation ... 3
      Spectacular Eclipse Experiences ... 4
   WONDERS SEEN IN A TELESCOPE ... 5
      The Appearance of Deep Sky Objects in Amateur Telescopes ... 7
   GETTING STARTED ... 8
      Planetariums ... 8
   ASTRONOMY CLUBS ... 9
      Getting Advice ... 9
      Sharing Knowledge and Ideas ... 10
   STARGAZING IS A MULTI-SENSORY EXPERIENCE ... 12

CHAPTER 2: THE BASIC BUILDING BLOCKS OF THE UNIVERSE ... 14
   STARS ... 15
   SOLAR SYSTEMS ... 15
   NEBULAS – INTERSTELLAR CLOUDS OF GAS AND DUST ... 16
   GALAXIES ... 16
   CLUSTERS OF GALAXIES ... 16
   SUPERCLUSTERS OF GALAXIES ... 17
   THE FILAMENTARY STRUCTURE OF GALAXY CLUSTERS ... 17
   EVERYTHING ORBITS SOMETHING ... 17
   EARTH'S LOCATION IN THE UNIVERSE ... 18

CHAPTER 3: UNFORGETTABLE STARGAZING EXPERIENCES ... 20
OBSERVING IN SURREAL LOCATIONS ... 21
Making the Most of Stargazing Experiences ... 21
Comfort ... 23
Two Unforgettable Comet Halley Events ... 24
Thrilled by Celestial Fireworks ... 27
Cruising Through the Galaxy ... 28
An Interstellar Canoe Ride ... 30
Dining Under a Heavenly Orb ... 31
The Appearance of the Full Moon Rising over a Perfect Horizon ... 31
Our Main Course with a Bronze Orb Suspended in the Milky Way ... 33
STARGAZERS GET TO SEE NATURE'S NEONS ... 34
Bioluminescence ... 34
Fireflies ... 37
Invasion from Space ... 38
Glow-Worms ... 39
Bioluminescent Plankton ... 39
Earth Lights ... 41
Auroras ... 42
Spectacular Lightning Displays ... 42
Coronal Glows ... 51
Fireballs ... 51
St Elmo's Fire ... 51
UNFORGETTABLE EXPERIENCES VISITING MAJOR OBSERVATORIES ... 52

CHAPTER 4: WHAT WONDERS CAN WE SEE IN THE SKY? ... 54
WHAT THE NAKED EYE CAN SEE ... 55
OBSERVING WITH BINOCULARS ... 57
WHAT TELESCOPES REVEAL ... 58

CHAPTER 5: ASTRONOMICAL NUMBERS, A LIGHT YEAR, STAR MAPS & ASTRONOMICAL TERMS ... 62
PART 1
ASTRONOMICAL NUMBERS ... 63

    A Light Year 65

    The Speed of Light 65

        The Speed of Light in Different Mediums 65

            The Distance Light Travels In 1 Second 66

PART 2
PRACTICAL TIPS FOR STARGAZING 66

    Navigating Your Way Around the Sky 66

    Star Atlases 68

    Measuring the Sky in Degrees 70

PART 3
COMMONLY USED ASTRONOMICAL TERMS 70

    Magnitude – a Scale for Measuring Brightness 70

    Integrated Magnitude 71

    Apparent Magnitude 71

    Intrinsic Brightness 72

    Apparent Diameter 72

    The Electromagnetic Spectrum 73

    A Binary Star System 73

    An Astronomical Unit 73

    An Astronomical Unit (AU) is the distance from the Sun to the Earth. 73

    Constellations 73

    The Celestial Poles 75

        Locating the North Celestial Pole (NCP) 75

        Locating the South Celestial Pole (SCP) 76

    Right Ascension & Declination 77

    Diurnal Rotation 77

    Culmination 81

    The Zenith 81

    The Ecliptic 81

    The Zodiac 81

    The Equinoxes 81

    Solstice 81

    Conjunctions 82

- Elongation and Conjunction ... 82
- Perihelion - Aphelion ... 83
- Deep Sky Objects ... 83
  - Common Names for Deep Sky Objects ... 83
- Messier (M) Objects ... 83
- NGC & IC Objects ... 84
- The Dark of the Moon ... 84
- Terminator ... 85
- Albedo ... 85
- Dwarf Planets ... 85
- Meteors & Meteorites ... 85
- Occultations ... 86
- Transits ... 87
- Phases ... 87
- Refraction ... 87
- Twilight ... 89
- Scintillation ... 89
- North, South, East & West ... 89
- Airglow ... 90

CHAPTER 6: LET'S GET PRACTICAL ... 91
UNDERSTANDING THE ATMOSPHERE ... 92
- The Structure of the Atmosphere ... 93
  - The Troposphere ... 93
  - The Stratosphere ... 93
  - The Mesosphere ... 93
  - The Thermosphere ... 93
  - The Ionosphere ... 93
  - The Exosphere ... 94
- How Thick is the Atmosphere? ... 94
- Cloud Types ... 95
  - *What Is the Composition of the Air?* ... 95

When Does Good and Poor Seeing Occur? 97
Do Mountains Have the Best Observing Conditions? 97
Why are Most Large Professional Observatories Located on High Mountains? 99
Is a Perfectly Dark Sky Pitch Black? 102
Airglow 102
Can More Faint Stars Be Seen from a High Mountain? 102
Sky Brightness 102
PREPARATION FOR DEEP SKY OBSERVING AT A DARK SKY SITE 104
Comfort 104
Food and Beverages 104
Wind Protection 104
Dressing for the Occasion 105
Managing Dew 106
Protection from Rain 106
An Eyepiece Case 107
The Effects of Body Heat 107

CHAPTER 7: MAKING THE MOST OF YOUR EYES 109
THE ADVANTAGES OF VISUAL OBSERVATIONS 110
How to Enhance Your Observing Sessions 110
PERSEVERANCE PAYS BIG DIVIDENDS 110
HOW OUR EYES WORK 111
How our Eyes Focus 111
Testing your Vision 111
Seeing Fine Detail 112
Visual Acuity Test 112
Observing Faint Objects 113
Dark Adaption 113
Substances That Reduce Dark Adaption 114
Test the Power of Full Dark Adaption 114
Night Blindness 116
Torchlight 116
Stray Light 116

Averted Vision ... 117
Threshold Vision Test ... 118
Astigmatism Test ... 118
Floaters ... 120
As Eyes Age ... 120
   Cataracts ... 121
   Macular Degeneration ... 121
The Iris ... 121

CHAPTER 8: TIPS ON OBSERVING WITH THE NAKED EYE & BINOCULARS ... 123

NAKED EYE OBSERVING ... 124
OBSERVING WITH BINOCULARS ... 125
A Good Binocular Star Atlas ... 126
Binocular Parts ... 127
How Binoculars Work ... 127
   Testing the Advantages of 3D Binocular Vision ... 128
   The Finger-Touching Depth Perception Test ... 128
Focusing Binoculars For Perfect Vision ... 128
The Field of View ... 130
Binocular Sizes ... 130
Large Binoculars ... 132
Binocular Mounts for Comfortable Observing ... 132
A Rotating Binocular Chair ... 134
Test the Quality of Binoculars Before Buying ... 136

CHAPTER 9: USING AND BUYING TELESCOPES ... 138

THE WONDER OF THE UNIVERSE REVEALED ... 139
TYPES OF TELESCOPES ... 140
Refracting Telescopes ... 140
Newtonian Reflecting Telescopes ... 140
Schmidt-Cassegrain Reflecting Telescopes ... 140
BUYING A DREAM OR A NIGHTMARE? ... 142

THE BASICS OF A TELESCOPE **143**
Aperture 143
Optical Quality 144
Good Collimation 144
Collimating a Newtonian Telescope 144
Focal Ratio & Focal Length 145
Magnification 146
High Magnification 147
Eyepieces 147
Using an Occulting Bar 148
Filters for Visual Observing 148
Telescope Mounts 149
WHAT TO LOOK FOR WHEN BUYING A TELESCOPE **150**
A VIBRATION FREE MOUNT **152**
IS BIGGER BETTER? **153**
A Motorized Observing Chair with a Power-Assisted Elevating Seat 154
Simple Observer's Seats 154
Choosing a Good Site to Set Up 155

CHAPTER 10: ENHANCING ONE'S OBSERVATIONAL EXPERIENCE BY DRAWING ASTRONOMICAL OBJECTS 156
WHY MAKE DRAWINGS? 157
COMPARING VISUAL OBSERVING TO PHOTOGRAPHY 157
HOW TO SEE MAXIMUM DETAIL 158
HOW TO MAKE A DRAWING 159
To Make a Pencil Drawing 159
Sketching Sunspots 160
Sketching the Planets 161
Drawing Rotating Planets 161
Preparing an Outline for Drawing the Planets 161
Observing Mars 163
Observing Jupiter 163
Observing Saturn 164

    How to Draw Deep Sky Objects 165
        Achieving Good Proportionality 165
        Drawing Nebulous Objects 166
    How to Make Negative Versions of Sketches 171
    How to Make Colored Sketches 172
        How to Change a Negative Drawing to a Positive One 173
    Night Sky Paintings 175
    Recording Scientifically Valuable Observations 176
        Recording Observing Conditions 176
        Recording Observer Variables 177

CHAPTER 11: PHOTOGRAPHING THE HEAVENS 179
    THE BASICS 180
    PHOTOGRAPHING STAR TRAILS 181
    CAPTURING THE MILKY WAY 183
    PHOTOGRAPHING AURORAS 185
    CAPTURING COMETS 185
    PHOTOGRAPHING METEOR SHOWERS 186
    CAPTURE THE MOON IN COLOR 188
    PHOTOGRAPHING THE FULL MOON RISING OR SETTING 188
    PHOTOGRAPHING LUNAR DETAIL 189
    PRODUCING HIGHLY DETAILED IMAGES OF THE MOON AND PLANETS 192
    RECORDING FEATURES ON THE SUN 193
    DEEP SKY PHOTOGRAPHY 193
    NIGHTSCAPE PHOTOGRAPHY 197
    THE NIGHT SKY AS SEEN WITH TIMELAPSE VIDEO PHOTOGRAPHY 198
    PHOTOGRAPHING GALAXIES TO DISCOVER SUPERNOVAS 198

CHAPTER 12: OBSERVATORY DESIGNS FOR AMATEURS 199
    WHY BUILD AN OBSERVATORY? 200
    THE SIZE OF AN OBSERVATORY 200
    DESIGN OPTIONS 201
        Traditional Domes 201
            The Skyshed Pod Observatory 203

ROLL-OFF ROOF OBSERVATORIES ... 204
The Telescope Mount ... 206
THE ROLL-AWAY SHED ... 207
THE ADVANTAGES OF A *DUAL* ROLL-OFF ROOF OBSERVATORY ... 207
Operating in Slit Mode ... 211
Seeing the Northern or Southern Horizon ... 211
Wall & Roof Construction ... 211
Wind Proofing the Interior ... 211
Interior Design Finishes and Lighting ... 211
Design Variations for Quality Constructions ... 215
Astronomy Club House Observatories ... 215
A Commercial Double Observatory for Public Viewing Nights ... 216
THE SPLIT SHED ROLL-APART OBSERVATORY ... 218
COMMERCIAL ATTRACTION CONCEPTS UTILIZING DUAL ROLL-OFF ROOFS ... 219
Commercial Astrophotography Observatories ... 224

CHAPTER 13: THE WORLD'S LARGEST OBSERVATORIES ... 227
THE WORLD'S LARGEST OPTICAL TELESCOPES ... 228
Australia ... 228
Siding Spring – the Anglo-Australian Observatory (AAO) ... 228
The Canary Islands ... 230
The Observatorio Del Rogue De Los Muchachos ... 230
Chile ... 232
The European Southern Observatory (ESO) ... 232
Mauna Kea ... 237
The Observatories ... 239
Russia ... 242
South Africa ... 242
United States Of America ... 244
Mt Graham Observatory ... 244
Mt Hopkins Observatory ... 244
McDonald Observatory ... 245
Kitt Peak Observatory ... 245

    Mt Palomar Observatory ... 247
    Mt Wilson Observatory ... 247
FUTURE GIANT OPTICAL TELESCOPES ... 248
    The Large Synoptic Survey Telescope (LSST) ... 248
    The Giant Magellan Telescope (GMT) ... 249
    The 30 M Extremely Large Telescope ... 250
    Comparison of Primary Mirror Diameters on Large Telescopes ... 251
    Size Comparison of the World's Largest Telescopes ... 253
THE LARGEST OPTICAL TELESCOPES IN SPACE ... 254
THE WORLD'S LARGEST RADIO TELESCOPES ... 257

CHAPTER 14: PRESERVING THE NIGHT'S NATURAL BEAUTY ... 260
THE OVERSELLING OF OUTDOOR LIGHTING ... 263
WASTED LIGHT AND ENERGY ... 263
    Billboard and Signage Lighting ... 264
    Street Lighting ... 264
        Cut-Off Street Lighting Protects the Night Sky ... 264
    Garden and Parkland Lighting ... 268
    Shielded Light Fittings for Businesses, Community Centers, and Homes ... 270
    So-Called 'Security' Lighting ... 272
ANTI-POLLUTION LAWS ... 273
    Light Pollution Ordinances ... 273
OBTRUSIVE LIGHT INCREASES HEALTH RISKS ... 273
ECOLOGICAL DAMAGE FROM WASTE LIGHT ... 273
A WORLD AWASH WITH WASTE LIGHT ... 275

ACKNOWLEDGMENTS

I give special thanks to the following people who offered valuable suggestions and comments when reading through the entire manuscript. They were my wife, Deborah Kelly, Nick Parfitt, John O'Sullivan Greg Bock and James Bryan. Others graciously assisted with various chapters. They included my son, Tane Thompson, Olivia King, Terry Lovejoy, Robert Stanford, Joe Gjerek, Dan McDonald, Brandon Clift, and Charles Gower. Nicole Brooke and Rebecca Gjerek worked with me on producing many of the diagrams. Joe Gjerek designed the excellent volume covers based on my rough ideas and his wife Rebecca produced a number of the complex CGI illustrations under my direction.

I must thank the contribution that my high school physics teacher Mr Thompson (no relation) and my chemistry teacher Mr McAlpine made by instilling in me the importance of scientific methodology, and importantly, the thrill of discovery and how to go about making accurate observations and how to conduct useful experiments (which unfortunately, has now been largely removed from high school science by non-scientific public servants with the view of eliminating all risk). While my tertiary education was valuable, it did not spark as much excitement as those early years of discovering the 'Wow!' factor in scientific inquiry and observational astronomy. In my adult working years when I built my own businesses, I found that my self-education in the sciences to be very enlightening. I gained the ability to learn across many fields and to see the connections between all the sciences. This proved to be very valuable in writing the last volume of this book.

Over the decades since I was in secondary school, there have been many amateur and professional astronomers in many countries, too numerous to mention here, who have helped me learn about a very broad range of aspects of astronomy. Many supported my astronomical endeavors in various ways, so I thank them for the knowledge they gave me to allow me to write this book. My good friend and expert telescope maker Cliff Duncan taught me how to build telescopes, and he also built excellent telescopes for me to use. He challenged me to think deeply about the many subjects that we often discussed at length. There have been many amateur astronomers that I have enjoyed the company of when at dark sky astrocamps where we would observe deep sky objects long into the night. There were also those who assisted me in creating and testing the Supernova Search Charts. Many professional astronomers believed in the value of those charts for making early discoveries of supernovas, and in doing so, they gave me much encouragement to complete the project, for which I thank them.

I have had a lifetime observing in the field as well as at private and professional observatories, and of course, in my own observatory. Over my life, I have had numerous discussions late into the night with both amateur and professional astronomers who gave me a diversity of insights into many aspects of astronomy and science in general. With their knowledge and my love of all aspects of astronomy and its associated sciences, this culminated in me producing *STARGAZING* with its wide-ranging subjects. It is a great joy for me to be able to share this knowledge with you.

I must also thank the scientists, science journalists, documentary makers, and the authors of my numerous science books that excited to me to learn since I was a child old enough to read 'How & Why' books. It always seemed that I could not get enough science books and magazines, or watch enough documentaries to satisfy my lust for knowledge of the big picture of things. I am indebted to the writers and producers of those exciting educational works because they allowed me to gather a wealth of knowledge to build an enormously valuable library of information.

I must also give a very special thanks to the scientists, engineers, and staff that have worked at NASA and the ESA for the truly incredible work they have done to make seemingly endless discoveries. They have unlocked so many mysteries, and in doing so they have exposed truly magical wonders in our solar system and many more in deep space. As a teenager, I was glued to the TV for each of the live feeds from the Apollo astronauts as they explored the Moon. I felt like I was there with them! And over the decades, to have seen what has been discovered by every mission to all the planets and many asteroids and comets, as well as the Sun, has been far beyond our collective imagination.

I cannot go without saying what an indelible impression the Hubble Space Telescope has had on me. What it has revealed has been nothing short of mind-blowing. And for it to show us such extraordinary beauty in all aspects of the universe, has been an incredible joy. To be aware of the astounding discoveries that NASA and the ESA have made, and are still constantly making, makes me feel very privileged and lucky to have been alive in this wonderful age of discovery.

All the people mentioned above have taught me parts of the underlying concepts behind what drives the most basic laws that control everything in the universe. And what an education this has been! Time and again, I had to let go of the common, simplistic beliefs we have been taught to see the more complex realities that control our world and the universe. I hope STARGAZING might convey some of these concepts to you so that you too can enjoy the wonder and joy that astronomy and science generally has given me. If I have done a good job with STARGAZING, you might pass this excitement on to others.

I am so thankful that I have grown up in 'The Lucky Country' of Australia, with its wonderful natural environments, beautiful modern cities, leading-edge technology, and pollution-free environments and exceptionally clear skies and good weather. Having had the opportunity to have traveled the world several times, this has enabled me to enhance my education in science, and also to have met many leaders in science in different fields.

No one in the history of mankind has had the opportunity to be aware of all that we know today. This has occurred due to our rapid advancement in technology, which has led to all the amazing discoveries that just keep coming. We are indeed very lucky.

My greatest joy has been the opportunity to share the wonder of the universe with others and to have had the opportunity to learn from them. It is my hope that STARGAZING will allow you to do the same and also have much enjoyment in doing so.

Gregg D Thompson

Stargazers get magnificent views of the night sky when they go camping away from the light pollution of developed areas. Credit: Adrian Mascenon

ABOUT THE AUTHOR

It would be difficult to imagine anyone more interested in all fields of astronomy than Gregg Thompson. Right from his early childhood and throughout his life, he developed a strong interest in all forms of science, but astronomy excited him the most as it encompassed most of the other physical sciences, and it was the one where the most amazing discoveries were being made.

In his late teens, he built his first telescope. In his 20's, he used his love of innovative engineering to design and build a highly practical and versatile observing chair to mount his large binoculars. (See Chapter 8.) Similarly, his love of architecture led him to come up with a new design for amateur observatories. It was a split roll-off roof design that was much more versatile and practical than other types of observatories. (See Chapter 12.) And like most amateur astronomers, he coveted ever larger telescopes.

ASTRONOMICAL INTERESTS

Not content with merely looking around the sky, Gregg's ambition was to make scientifically-useful observations. Before the days of high resolution astrophotography, he applied his artistic talents to making accurate, detailed drawings of the planets, the Sun, and comets.

Gregg meticulously observed hundreds of galaxies searching for exploding stars (supernovas). He spent 12 years developing maps of star fields around 300 of the brightest galaxies. This allowed amateur astronomers to make visual discoveries of supernovas that randomly occur in galaxies. In 1989, Cambridge University Press UK published his work, which he co-authored with another amateur astronomer, James Bryan Jnr from the University of Texas. James observed galaxies around the North Celestial Pole that Gregg could not see and assisted with the editing of the book. The book was entitled '*The Supernova Search Charts and Handbook*'. For this work, Gregg was awarded the Amateur Achievement Award from the Astronomical Society of the Pacific in the US (the world's largest body of professional and advanced amateur astronomers). He shared the award with his colleague Rev Robert Evans who discovered many supernovas visually with the help of Gregg's charts. His charts also received the 'Highly Recommended' status from the prestigious *Rolex Awards for Excellence in Science*. Gregg's supernova search charts enabled amateur astronomers to discover bright supernovas, as they occurred. Most supernovas were previously discovered on photographs long after they exploded, so it was not possible to observe how they developed. By discovering supernovas as they started to explode, professional astronomers could then make detailed observations of these events using large telescopes and even the Hubble Space Telescope, thereby enabling astrophysicists to better understand the physics underlying each type of supernova. Supernova discoveries were critical to enable professional astronomers to determine the size and age of the universe. This led to the discovery that the rate of the universe's expansion was unexpectedly increasing. (See Volume 4 Chapter 1, Part 5.)

Gregg also produced at the same scale as the supernova charts 1,000 photographs of small faint galaxies taken from the 48" Schmidt plates that amateurs could use to visually detect supernovas in galaxies more distant than those in his charts.

Gregg also observed and recorded the visual appearances of hundreds of nebulas and star clusters in the Milky Way, as well as those in the Magellanic Clouds. For this, he used his 310 mm and 460 mm telescopes.

In 1992, the editor of Australia's largest publishing house, Weldon Publishing, attended one of Gregg's themed, special-effects stargazing shows for guests at a major Australian country resort, Coolum Hyatt. Many people from outside the resort traveled long distances to attend his shows. The resort is located in a dark sky location. The editor was impressed with Gregg's knowledge of the night sky and the way he conveyed his enthusiasm for astronomy, as he showed his thrilled audience objects in space in his large telescopes. She liked the exciting way in which he presented facts about the universe complete with atmospheric special effects lighting and a very large, specially-constructed screen on which he projected video vision of flying past the planets and into objects in deep space. She thought his fibre-optic lit Star Trek-like control center and his spacey sound effects added a lot to the atmosphere of the evening's experience. In view of this, she asked him to write a popular book on observational astronomy. It was titled '***The Australian Guide to Stargazing***'. It was published in 1993. To his amazement, the book was widely lauded and made the top-seller list. With new additions and reprints it has sold consistently for 25 years.

Gregg's lifelong commitment to astronomy brought him into contact with many novice and advanced amateur astronomers, as well as professional astronomers, astrophysicists, and those in the media who promote astronomy. He organized, and participated in, many star parties at dark sky locations, as well as orchestrating tours of the night sky for members of the public at eco-resorts, and for guests on cruise ships on which he lectured. Gregg has been an invited guest speaker at astronomy conferences, club meetings, and business functions. He has taught educational courses in astronomy, and he has had many groups and schools visit his observatory. He has regularly been a guest on talk-back radio shows and TV programs where he would discuss a broad diversity of astronomical subjects and what was current phenomena to observe in the sky. He also spoke on how to control light pollution and he would explain what perceived UFO sightings actually were. Gregg's intellectual contribution to astronomy has been well-recognized.

BUSINESS PROJECTS RELATED TO ASTRONOMY

In business, Gregg is an accomplished entrepreneur. He has served as the managing director of a number of companies he has founded. One of Gregg's companies created promotions for national and multinational companies. Some of these have been based around astronomical themes or phenomena.

For the return of Halley's Comet in 1986, Caltex Oil (a division of Exxon) sponsored very successful educational charts and planispheres for this event. Gregg designed, produced, and successfully marketed these in Australia, New Zealand, the USA, Hong Kong, and Japan.

THE INFINITY EXPERIENTIAL IMMERSIVE MAZE

For another of his companies, Dreamtech Designs & Productions, Gregg designs, constructs, and operates multi-million-dollar themed leisure and edutainment attractions. Gregg's highly-acclaimed, experiential INFINITY attraction features a journey through a series of 20 multi-sensual, fun-filled, immersive

Right: '**The Supernova Search Charts and Handbook**' provide a means for amateur astronomers to discover exploding stars in hundreds of distant galaxies.

Left: '**The Australian Guide to Stargazing**' was a best seller.

art environments, which appear to recede to infinity. This 21st C funhouse features innovative special effects in immersive art environments. Its illusions push the boundaries of experiential themed attractions to the limit of technology. For instance, in the peaceful Star Chamber, patrons walk into heaven at the center of the galaxy where millions of stars extend for as far as the eye can see. Angelic choral voices make one feel like they are in heaven. In another environment, people laugh and dance on a waveform floor in the dark with only the light of colored starlight and their hands and feet glowing. In an endless disco-like environment, large spheres appear and disappear as they are bounce around in slow motion to ultra-groovy music. INFINITY's environments elicit a broad range of emotions. This has made it a successful attraction that is very popular for all ages.

THE SPACEWALKER EDUTAINMENT EXPERIENCE

In 2005, Gregg also created the *SpaceWalker* edu-tainment attraction, which allows visitors to discover the universe in many unique ways. This science-based attraction creates a high level of visitor interaction and participation. It utilizes leading edge special effects and highly themed environments that are suggestive of popular science fiction movies to give visitors many fun experiences, as they learn remarkable things about the cosmos.

As examples of his command of detail in the solar system section of the 'Space Walk Highway', Gregg created very realistic models of the planets, which appear to be floating in space as they rotate! Even special effects experts were not able to work out how he achieved this. He did not use holograms but real models. Moons orbit the Gas Giant planets with no apparent means of support. They cast their shadows on their planet's cloud tops, as they transit their planet's globe. Patrons can observe comets, asteroids, and space probes traveling through the solar system.

As Space Walkers journey through the spiral arms of our Milky Way galaxy, they see faithful, three-dimensional recreations of all types of deep sky objects. Gregg developed a means of constructing a three-dimensional globular star cluster with thousands of stars that appear to be suspended in space. Patrons pass by exo-planets orbiting other stars, as well as rapidly rotating neutron stars, and matter spiraling into a black holes. A supergiant star can be seen losing its outer atmosphere to a brilliant white, super dense, dwarf star in orbit around it.

After leaving our galaxy, Space Walkers travel into the depths of intergalactic space where they observe clusters of galaxies in every orientation that extend to the edge of the observable universe. Here, they can enter a rotating black hole where they travel through time. Or, they can take a return journey to the space station via a monster star gate filled with stars. Halfway through, Mission Control advises that aliens have invaded the *Stargate* as it drops into darkness. Patrons have to avoid being eaten by the monsters from the movie '**Alien**'.

In either case, visitors emerge safely at the entrance of the futuristic Restaurant at the End of the Universe to experience creative taste sensations from across the galaxy. Food and drinks are presented in very original ways where they light up and change color.

Many visitors to SpaceWalker commented on the extraordinary attention to detail and the imaginative concepts used. Images have been included to help readers appreciate the level of detail, the innovation, and the powerful means of communicating science to the public that Gregg has used. He has a talent for making learning about the universe exciting.

Gregg is a polymath due to his love for all the sciences – physics, chemistry, geology, meteorology, biology, neurology, psychology, sociology, logic, and of course astronomy. And because the universe includes life, astronomy indirectly includes the sciences that deal with life, us, and intelligence.

Over decades, Gregg has had considerable opportunity to discuss all manner of subjects related to astronomy with numerous people from all walks of life. This has given him a great appreciation for the sorts of subjects that interest amateur astronomers and general readers who have a broad interest in astronomy. He has included a wide diversity of subjects related to astronomy throughout STARGAZING, and especially in the last chapter.

When Gregg shows city dwellers the stars under a naturally dark sky, they are typically amazed at the vast number they can see. They are also surprised by how many meteors and satellites are visible. They are impressed with the easy visibility of the Milky Way and its delicate beauty. Most people today have never seen the Milky Way due to light pollution. It's only under dark skies that one gets to see the elusive Zodiacal Light and many faint meteors.

Gregg describes the effect that astronomy has had on his life by saying *"Once the doors to our imagination and curiosity are opened, we discover that the universe overflows with wonder, beauty, and mystery."*

Astronomy has given Gregg much enjoyment throughout his life, so it is his hope that he can repay what he has gained by sharing what he has learnt and experienced with *STARGAZING's* readers. They too will gain the sense of awe and wonder that he has experienced throughout his quest to discover the marvels of the universe.

Gregg thinks that life is going too fast and that in time, he will run out of years, so before that happens, he wanted to use all he had gained from his life in astronomy to write the ultimate book on stargazing for those interested in observing, and learning about, all aspects of the universe. As he has no need to profit from this project (that has taken him seven years to complete) he wanted to sell it at a low price, even though this would be unlikely to recover the costs involved in producing it, and then publishing it, let alone his time. He has done this so that it would be easily affordable to as many people as possible. He did not want to take this knowledge to the grave, so he hopes readers will agree that he has made the right decision.

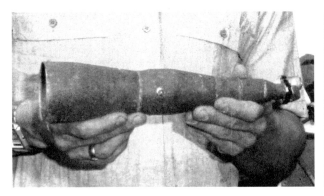  

Left: **Gregg's father bought him his first small telescope when he was 11. It was a very well used, terrestrial spotting telescope with a 60 mm (2.5") diameter lens and a magnification of 30 times. What he saw through it captivated his imagination so much that he became very enthusiastic about learning all he could about astronomy. He was also given the Larousse Encyclopedia of Astronomy which enthused his to want to see everything in that large book.**

Center: **By his early teens he had upgraded to a 60 mm Unitron refractor that permitted him to use magnifications up to 300 times. It also had very convenient slow-motion controls and a view finder.**

Right: **In his mid-teens, he worked through his school holidays to purchase a 114 mm (4") Royal reflector.**

Left: **In his late teens under the guidance of a beloved mentor, he learnt to grind and figure a 200 mm (8") f 6 mirror for a Newtonian reflector, which he mounted on a car axle.**

Right: **In his 20s, he used a 200 mm (8") f8 high resolution reflector telescope, which was made by his close friend and expert amateur telescope maker, Cliff Duncan who figured many mirrors to perfection. Cliff constructed his own foundry to make specialized parts for the telescopes that he generously designed and built for Gregg and other amateur astronomers at no cost - just for his love of the art of telescope making.**

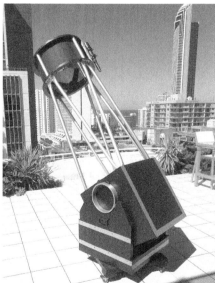

Left: **In Gregg's 30s, he acquired a German mounted, high resolution 310 mm (12.5") Newtonian reflector.**

Center: **A decade later he acquired a 460 mm (18") high resolution Dobsonian reflector.**

Right: **Throughout those years, he used his very practical binocular chair for observing with his large 20 x 80 binoculars.**

INFINITY Experiential Illusion Maze

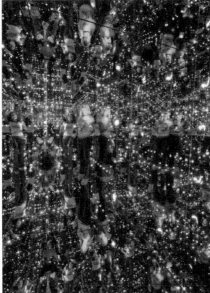

Left: **In Infinity's 'Infinite Kaleidoscope', people dance to music while immersed in a continuously changing kaleidoscope of colorful images that extend for 20 stories in every direction.**

Center: **Infinity's 'Electron Maze' simulates being inside the 'mind' of a gigantic computer where electrons are racing along circuits that never end in every direction. Each circuit has its own color and sound field. Patrons have to find their way out of this seemingly endless maze to get to their next experience.**

Right: **Patrons dare to cross Infinity's 'Light Canyon's' suspension bridge, which extends over a bottomless chasm.**

An infinite computer-like maze **An infinite light chamber** **An infinite disco of disappearing**

SpaceWalker Edutainment Attraction

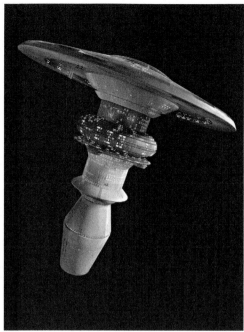

At Earth Station, visitors experience a simulation of being dematerialized and teleported to the giant Space Station Zeta out in the galaxy. Here, advanced aliens enlighten humans about amazing aspects of the universe.

Left: **The Arrivals Centre** *at* **SpaceWalker's Star Station Zeta**

Right: **This is Pod 1 where audiences are taken on a virtual reality flight over Venus and Mars's extraordinary landscapes.**

Left: **In one of SpaceWalker's Discovery Pods, humans are suspended in a 'mind transfer' unit that recreates a pleasant evening on Earth.** An alien spacecraft descends from the stars to take Earthlings on a mind-bending, faster-than-light journey to the edge of the observable universe so they can learn about the structure of the cosmos.

Right: **At the end of the universe, visitors travel through the spinning accretion disc of a black hole to go inside a space warp that transports them back to the space station's Restaurant at the End of the Universe.**

Left: **At the outer edge of the space station, spacewalkers can pass through an airlock into a rotating Stellar Vortex Wormhole (right).**

Left: **Here we see Discovery Pod 2, which takes spacewalkers on journeys to strange but real planets.**

Right: **At edge of the space station there is a Stargate where patrons take a leap of faith to jump into to experience micro gravity amongst a sea of stars in every direction.**

The Space Walk section realistically replicates a superb diversity of all types of objects in the universe. Travel along a 'highway through the universe' that goes through our solar system, then into the Milky Way galaxy, and finally into intergalactic space. Gregg created highly atmospheric sound fields to give each section a very 'spacey' feel to it. Along this highway are 'space-time viewers' in which visitors can watch exciting 1 to 2 minute videos of the most impressive features of each type of object in the universe. Gregg is justifiably proud of the quality of the three-dimensional models that he and his artisans created under his guidance. Each object was meticulously crafted to realistically replicate their real-world counterpart.

This extraordinarily accurate 3D model of the Moon took a master sculptor six months to sculpt all features in relief. Each of the Moon's numerous features had to be located exactly to perfectly replicate the detail seen in small telescopes. As one walks along the Space Walk highway, the Moon can be viewed from New Moon through to Full Moon, and due to its slow rotation, the far side of the Moon is also visible.

Left: **Jupiter and Saturn** were very difficult to create because they had to be perfect oblate spheroids with their moons orbiting them at different rates. Jupiter's Galilean moons had to pass in front of the planet with no apparent means of support. Each moon cast its shadow on Jupiter's cloud tops - just as they do in the real universe.

Right: **Saturn** was especially difficult to light due to the gap between the rings and the planet's globe. The rings had to cast their shadow on the globe while the globe cast its shadow on the back of the rings, yet not spill any (sun)light. Like the real planet, the rings rotated, and Saturn's family of moons orbited the planet at different rates.

AN INTRODUCTION TO THE STARGAZING VOLUMES

When talking to amateur astronomers from all over the world, over the decades I often heard calls for a comprehensive book that covers all aspects of stargazing. I believe the 4 Volumes of *STARGAZING* satisfy that need because they incorporate the following:

- All the **practical things you need to know to stargaze** for the joy of it as well as doing **serious observational astronomy**,

- Learning the basics about **how the atmosphere works** to maximize one's opportunity of observing under the best conditions,

- **Understanding how your eyes work to maximize our vision** to see very faint objects and to detect maximum detail, as well as knowing **how to test your eyesight**,

- **How to prepare for observing sessions**,

- **The advantages of observing under a dark sky**,

- **Which objects are best for viewing with the naked eye, binoculars, and telescopes**,

- Tips on **buying and using telescopes**, and other astronomical equipment,

- The types of **observatory designs** that are the most practical,

- **How to minimize light pollution**,

- **Finding one's way around the sky and understanding astronomical terms**,

- **Understanding the nature and the significance of what you are observing**,

- How to get started with **astrophotography**,

- How to go about **observing the Moon, the Sun, the planets, asteroids, and comets**,

- **Observing the best examples of all types of deep sky objects** (star clusters, nebulas and galaxies) in both the northern and southern hemispheres,

- Observing all types of **astronomical events** such as **eclipses** of the Sun and the Moon, **meteor showers**, **satellites**, **auroras**, and the **Zodical light**,

- Descriptions of the **world's major observatories** for stargazers to visit,

- **Drawings of astronomical objects** to illustrate what objects look like visually, and **how to make your own quality drawings**,

- **Excellent information graphics** for concepts, types of objects, and astronomical events,

- The **best and latest images** taken by large **professional observatories, space telescopes and space probes**, as well as those taken by **the most advanced amateur astronomers**. Impressive 3D images are included,

- **The best space art** by the world's leading CGI artists,

- The **latest discoveries** across each field of astronomy and related sciences with **easy-to-understand explanations**,

- **Personal anecdotes** about unusual observations and events,

- How to have exciting **'Wow!' experiences** when observing,

- **Observing challenges** that stargazers like to test their observational abilities,

- **A big picture view of the size, structure, and the evolution of the universe,**

- The likelihood of **life and intelligence** in the universe,

- **The latest concepts and discoveries about cosmology,**

- **The most remarkable aspects of the history of astronomy** so that readers can appreciate how much we owe to our greatest, intuitive visionaries,

- **Answers to the BIG questions** that people want to know that astronomers, physicists and cosmologists are discovering.

Because *STARGAZING* is so all-inclusive, it took seven years to write it and edit it many times over. Exhaustive research was done to ensure it was up-to-date with the latest discoveries. Creating many of the illustrations as well as finding and collecting all the images were also massive undertakings. Much time was consumed in managing numerous communications, preparing the layout for each chapter and creating the website. When all this was done, organizing the distribution of the book and its marketing were also major tasks.

Because the book is a large body of work that has a high content of superb, large color images, it would be prohibitively expensive to publish it as a traditional printed book. This would severely limit its potential readership. In view of this, I decided to make it an ebook to make the cost minimal so it would be appealing to a broad audience. This would allow me to share all that I have gained from astronomy with as many readers as possible.

STARGAZING'S EXTRAORDINARY IMAGERY

The book's page size was originally designed as a 260 mm x 300 mm (10" x 12") printed coffee table book, but as the book grew in size, it had to become an e-book. This allowed very large images to be included so they can be zoomed into to see remarkable detail.

Thanks to recent advances in digital photography and computer processing, amateur astronomers are now capturing excellent, high-resolution images, even of very faint objects. Because of this, *STARGAZING* features the best **astrophotography taken by the world's most advanced amateur astrophotographers** from many countries. These images have a depth of color and detail that was unimaginable even a decade ago.

There are also stunning images taken by **large, professional, ground-based observatories, space telescopes, and space probes**. The very high-resolution images taken by the Hubble Space Telescope (HST) and the European Southern Observatory (ESO) are simply stunning.

The book's imagery also includes reasonably scientifically accurate, **space art** painted by a number of the world's leading space artists. Their illustrations allow readers to visualize what planetary landscapes, exo-planets, bizarre stars, and other solar systems might look like.

STARGAZING also features **drawings of astronomical objects** made at the telescope in both black and white, and in color. These drawings illustrate what objects appear like visually so that novices do not expect to see Hubble Space Telescope images in their telescopes.

STARGAZING's images convey far more information than written words could ever do, and they demonstrate the extraordinary beauty that abounds across the universe.

WHO IS STARGAZING WRITTEN FOR?

STARGAZING's volumes are written for anyone interested in astronomy:

- **novice stargazers** who want to know how to go about observing and what exciting things they can see,

- **experienced amateurs** who want to know more about the physics of the objects and the phenomena they are observe,

- **advanced amateurs** who want to broaden their general knowledge into areas that they have not delved into,

- **professional astronomers and many scientists** who are so consumed in their field of expertise that they find it hard to keep up with discoveries beyond their specialty. The latest concise discoveries in STARGAZING are appealing to all readers,

- **astrophotographers** who love to see the most extraordinary imagery in every field of astrophotography,

- **students and science teachers** who find STARGAZING especially good for learning about all aspects of astronomy and its associated sciences,

- **general readers** who love learning about the cutting edge of astronomy and cosmology,

- **futurists** who want to visualize what the near and far future will be like, and the role that technology, intelligence, and consciousness may play in the evolution of the universe.

STARGAZING'S MAGIC MIX OF SUBJECTS

STARGAZING contains **exciting explanations of many subjects**, together with excellent **info-graphics** about **the characteristics of all types of astronomical bodies and the diversity of phenomena in the universe.** The text describes how such things work and evolve in **easy-to-understand language**. As well, it has numerous **thought-provoking concepts** supported by **helpful analogies** that make them easier to understand. The book is interspersed with surprising historical facts, and amusing personal anecdotes. Where applicable, it also includes references to **science fiction movies and documentaries** that are related to the subjects being discussed.

I am very lucky that astronomy has given me more than my fair share of 'Wow!' experiences, particularly when I have been observing at very atmospheric locations and when witnessing awe-inspiring astronomical phenomena. I have shared these experiences throughout the book, so that others might delight in having similar experiences.

Most readers should find *STARGAZING*'s magic mix of diverse subjects combined with its exciting style of presentation, to be a thrilling and pleasurable read.

WHAT YOU SHOULD KNOW ABOUT THE STARGAZING TEXT:

- It requires **no specialist knowledge of science or mathematics,**

- **Names, technical terms, and important words are shown in bold** so they can be located quickly when scanning the text.

- Many sections have **summaries of a section shown in light blue,**

- The author's **anecdotes are shown in navy blue,**

- **Challenging questions** for readers to contemplate are shown in violet. When there is an observational challenge, this is indicated by the icon Ⓒ. These challenges are given because many amateur astronomers like to test their observational skills and their instruments,

- Throughout the text, are **information 'boxes'.** These have a blue-gray background. These are included to provide additional interesting information that is indirectly related to the topic of discussion,

- **Most quantitative information is presented as rounded off numbers** because many measures and quantities in astronomy are estimates. Rounding off numbers makes them easier to remember,

- **Measurements** are given in Metric units with Imperial units following in brackets. Kilometers are abbreviated to **km**, meters to **m**, millimeters to **mm**, miles to **mi**. Feet are denoted by ', and inches by ". Minutes of arc are denoted by ' and seconds of arc as ". Kilograms are abbreviated to kg, and pounds to lb. Temperature is abbreviated to C for Centigrade, F for Fahrenheit, and **K** for Kelvin,

- **Major well-known professional telescopes, observatories and Space Agencies** are abbreviated as follows: Hubble Space Telescope as **HST**, National Aeronautics and Space Administration as **NASA**, the European Space Agency as **ESA**, the Japan Aerospace Exploration Agency as **JAXA**, the European Southern Observatory as **ESO**, and the National Optical Astronomy Observatory as **NOAO**.

- **Star charts** and **Celestial coordinates** of deep sky objects are not provided because they are seldom required nowadays due to most modern amateur telescopes using computer-aided object finding technology (CAT systems). If coordinates are required for 'fixed' objects, such as stars, nebulas, and galaxies, they can be found using night sky Apps or a smartphone. They are easy to find by going on the Internet and typing in the object's name. Coordinates for moving objects such as comets and asteroids can be found online by searching for the body's name and appending the word 'coordinates'.

In the vein of Carl Sagan's extraordinarily popular 'Cosmos' series, STARGAZING includes some **thought-provoking speculation** about many subjects. The speculation in STARGAZING is founded on recent scientific discoveries and through making logical deductions.

The rate at which major discoveries in all sciences are racing ahead, is forcing us to rethink many things that that we thought were solid facts and commonsense. In some cases, we are now having to question what is reality. As we learn more with an open mind, we have to be prepared to let go of old, simplistic beliefs and consider the evidence that exists for some very complex aspects of nature. Some of the conclusions that we are confronted with seem impossible at first, but once understood, they are intellectually very stimulating and they give us a new view of nature. The evidence that supports them makes them hard to refute. This is particularly the case for many subjects in Volume 4.

Since I was a child, I have had a lifelong love for astronomy due to the awe that it has inspired in me. Over the years that STARGAZING has taken me to compile it, I have learned a great deal about many mind-expanding discoveries that are at the leading edge of astronomy, and I have become aware of amazing new technologies that will dramatically change our future. As well, I have discovered truly incredible aspects about the subjects that are involved in the evolution of our universe. Putting STARGAZING together has been a very exciting and stimulating cerebral journey for me. As you read through it, you too will embark on this amazing journey. Not only will it open your mind to incredible concepts, you will see unbridled beauty. You will also time-travel back to the very beginning of the universe's creation and into its distant future!

I hope you will enjoy the 'ride' you are about to take and the knowledge that you will gain as much as I have.

Gregg D Thompson

Chapter 1
THE WONDER AND JOY OF STARGAZING

Stargazers use their telescopes to peer deep into the universe to see awe-inspiring sights. Here we see a group of amateur astronomers observing at a dark sky site away from light pollution at a star party in Texas USA. Credit: Alan Dyer

WHY IS STARGAZING SO EXCITING TO SO MANY PEOPLE?

Because astronomy has so many aspects that capture people's imagination, it is no wonder it is the most popular of the sciences. It is fairly easy to relate to what is in the universe without needing to have a lot of prior knowledge of astronomy. People are able to share in the excitement of what they can see in space and we are continually amazed by the constant stream of unexpected, and often astonishing discoveries.

Astronomy began when ancient people first noticed how the stars moved across the sky. They gave names to different groups of stars, which we now call the constellations. They found that when certain constellations appeared in the early evening sky, they foretold of the coming of the next season. There were no clocks and no calendars back then to keep track of time, so the stars were a convenient method to do so. The earliest primitive astronomers became aware that some bright stars wandered between the fixed stars and that they changed in brightness over time. The 'wanderers' were thought to be heavenly gods with supernatural powers. Today, we know them as the planets. Over many millennia, shaman attributed many myths to them and this became known as astrology. For thousands of years, it was commonly believed by those with no knowledge of science that the stars influenced people's lives and that they determined their personalities. Many thought they could even foretell the future. But when these beliefs were tested, they were found to be no more accurate than guessing. Genetics has shown us that our personalities are controlled by our genes, and in some cases, by a major event in our lives.

In the 16th C, **Galileo's** newly-invented telescope that used a lens was able to gather a lot more light than the naked eye could, so this made faint objects in the night sky visible. His amazing invention could also magnify objects thereby allowing more detail to be seen.

Sir Isaac Newton invented a new type of telescope in the 17th C that used a mirror. It was considerably more powerful than Galileo's. Newton invented the 'reflector' telescope. It used a larger diameter mirror that was part of a sphere to gather much more light than lensed telescopes could do at that time. Newton's small reflector telescope has evolved into today's high-tech monsters. (See Chapter 13.) These are revealing stunning detail that not long ago was not even conceived as being possible. In recent decades, reflector telescopes on space probes have revealed unexpected detail on worlds right across the solar system, as well as the most stunning, up-close images of our Sun showing its explosive outbursts. (See Volume 2 Chapter 3.) Telescopes in space have produced such as the extremely successful **Hubble Space Telescope (HST)** has made numerous surprising discoveries and captured the most astounding images that we could never have imagined seeing before it was launched. It has revealed exceptional beauty in numerous star clusters, gaseous nebulas, and in the diverse forms that galaxies have. We can now see right across the universe back to the beginning of time when the universe was created!

Recent unexpected astronomical discoveries about other worlds have become stranger than science fiction. (See Volume 2.) Imaginative science fiction writers are now able to use concepts based on reality for their stories, rather than fantasy when they envisage strange planets. Strong public interest in ground-breaking discoveries and what lies beyond our planet, is driving the production of many exciting documentaries and movies about outer space. There has never been a time that has made astronomy more exciting than now.

Thousands of planets are now being discovered orbiting other stars. Our knowledge of other worlds is now growing exponentially. When combined with an understanding of how life evolves, we are seeing that life is probably very common throughout the universe because it appears to be a natural outcome of the universe's evolution. It is highly likely that life may have evolved in many solar systems across the universe billions of years before it evolved on Earth. And more than likely, we are not the only intelligence in the universe. It now appears that the universe may be filled with intelligences far more advanced than us late-comers.

We are now discovering the universe's size, its surprising structure, its evolution, its age and how it is likely to die. Cosmologists are trying to understand how all its energy and matter could have come from apparently nothing when the Big Bang occurred. Some are wondering if there may be other universes. (See Volume 4 Chapter 1.)

STARGAZING'S 'WOW!' FACTOR

It's a truly moving experience to look up into the night sky's vast panorama when you have an unobstructed view on a clear, dark, moonless night, far away from urban light pollution. It's especially magical when the Milky Way stretches right across the heavens from horizon to horizon. Most people who live in cities have never seen this awe-inspiring sight. There are thousands of stars so crisp and numerous that many people say they feel like they could reach up and touch them! If you recline in comfort in a sun lounge or on a ground sheet and gaze straight up into the heavens above, you feel like you are immersed in the universe's vastness. Sometimes an unexpected brilliant meteor will startle you as it streaks across the sky. Some are bright enough to light up the whole landscape for second or two. The brightest will leave behind a trail of light for a minute or more. In *STARGAZING*, there are many examples of observing experiences that are so memorable that they often create emotive bonds with those with whom you share these experiences.

The settings in which stargazers are often immersed can make their experiences very special and romantic. Following are some examples.

TEN-THOUSAND STAR ACCOMMODATION

Camping with friends or family under the stars at a dark site can be a marvelous and memorable experience, especially if you know what you are looking at in the starry vault overhead. You'll see the wonder of our galaxy and the bright naked eye planets. Each has its own particular color. You will notice that the stars are not all white, some of the brighter ones are orange, blue, and yellow. In a telescope, you will be able to observe a wide variety of hundreds of star clusters, nebulas, and galaxies in some detail. The largest and brightest can be glimpsed with the naked eye and in binoculars. From time to time, bright, naked eye comets with their long tails can be viewed with the naked eye appearing to

This is a typical, awe-inspiring naked eye view of the night sky when observed from dark country locations.

> *In this marvelous photo taken in the Chilean Andes taken by Miloslav Druckmuller Silhouetted along the spiral arms of the Milky Way, we see dark clouds of dust and gas that are the dust lanes of our galaxy. Stars and planets condense from this matter. Nebulas and star clusters can be seen with the naked eye as hazy spots. To the upper right, we see what looks like two detached portions of the Milky Way. These are two dwarf galaxies that orbit our Milky Way. They are known as the Large and the Small Clouds of Magellan. A short meteor trail is visible to the lower left. Credit: Miloslav Druckmuller*

You'll be wowed when you see a bright, bolide lighting up the sky as well as the surrounding countryside. Credit: Howard Edin - taken at the Okie Tex Star Party

It was awe-inspiring to look at a large, bright naked eye comet Like Comet Mc Naught with its wispy tail sprayed across the constellations. This photograph was taken from Swift's Creek, Victoria, Australia on 23 January 2007. Credit: Fir0002 / Flagstaffotos / Wikimedia Commons CC BY-SA 3.0.

hang motionless amongst the stars. Telescopic comets are frequent visitors to the inner solar system.

If you are a city dweller, to see the sky at its very best, it is worth driving for 30-60 minutes into the country or to a National Park to get away from the city light pollution, so you can see the beauty of a naturally dark sky. However, the Moon, the planets, and bright star clusters can be well seen from urban locations.

Astronomy encourages enthralling conversations with other observers, especially when relaxing with a beverage after an observing session. I remember on one occasion at midnight, friends and I were nestled amongst large boulders on top of a spectacular, remote, 60 million-year-old extinct volcano. We had distant views over an expansive plain stretching below us for as far as the eye could see. Sprinkled across it were scattered dim lights from country ranches and small townships. While gazing into the star clouds of our galaxy, we chatted into the wee hours of the morning delving into the possibility of life out there and what it might be like.

SPECTACULAR ECLIPSE EXPERIENCES

Many people who experience a **Total Solar Eclipse** for the first time, are so in awe of this phenomenon that they travel the world to witness other eclipses. When a total eclipse is viewed in a clear, it would have to rate as one of the most impressive wonders of nature. (See Volume 2 Chapter 3.)

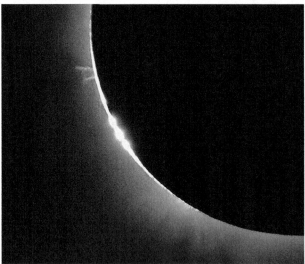

Top: **Here we see the last glint of sunlight shining through a valley on the Moon during the commencement of a total solar eclipse. This also occurs at the end of totality. It is known as the Diamond Ring effect. In binoculars and a telescope, gigantic, hot pink flames (prominences) can be seen leaping out of the Sun's limb. The largest can be seen with the naked eye.** Credit: Jason and Edwin Aquirre

Bottom: **A prominence becomes visible in binoculars just as the bright surface of the Sun disappears behind the Moon;' limb.** Credit: Jamie Vilinga

Many astronomical scenes can be quite surreal like this total lunar eclipse occurring over a town shrouded in haze. Credit: C Haney

During a total lunar eclipse, the Moon becomes a deep bronze color.

When the Moon goes into the Earth's shadow, a **Total Lunar Eclipse** occurs. When this happens in the early evening, it is a good time to have an eclipse party. (See Chapter 3, page 24 to see how to make this event a very special occasion.)

WONDERS SEEN IN A TELESCOPE

It is exciting when you first see the incredible detail that is visible on our Moon's dramatic landscape. Even a good quality, small telescope will show striking detail in the Moon's mountain ranges, its large plains of ancient lava, and its enormous craters. Each night new lunar features are visible. In large amateur telescopes, they are simply stunning. (See Volume 2 Chapter 5 for many examples.)

The Moon presents an amazing degree of detail when viewed in amateur telescopes. Credit: Kitt Peak Observatory

There are impressive sights to be seen in amateur telescopes when observing the planets Jupiter, Saturn, and Mars.

Left: **The planets Jupiter, Saturn, and Mars, are enthralling to observe because they are continually changing. It's exciting to watch Jupiter's storms continually changing and its four major moons passing one another and casting ink-black eclipse shadows on Jupiter's cloud tops.** (**See** Volume 2 Chapter 9.) Credit: Faulkes Telescope Project

Center: **To see Saturn suspended in space with its rings and moons orbiting it, is a moving experience. Over 7 years the rings change from edge-on for a day to open to their widest extent. Each night you can watch its moons move around the planet as if it is a mini solar system.** (**See** Volume 2 Chapter 10.) Credit: HST

Right: **When Mars is close to Earth, it is captivating to see it rotating before your eyes, and watching its seasons change over a few months.** (**See** Volume 2 Chapter 8.) Credit: NASA

THE APPEARANCE OF DEEP SKY OBJECTS IN AMATEUR TELESCOPES

Left: **It's delightful to see an open cluster of bright colored stars of different brightnesses suspended in space.** (See Volume 3 Chapter 6.) Credit: Steve Crouch

Right: **To peer into a wide-field eyepiece and see a distant globular cluster tens of thousands of light years away containing thousands of stars peppered across a spherical glow of a million more stars too faint to be resolved, is a breathtaking sight.** (See Volume 3 Chapter 7.) Credit: ESO

Left: **When you see your first planetary nebula of gas that has been thrown off by a dying star, it is a real joy. This image simulates a view at medium power.** (See Volume 3 Chapter 5.)

Center: **When you see a large, diffuse nebula filled with star birth, as well as dark dust clouds silhouetted against glowing gas, this is something to behold. This view of the Eta Carina Nebula simulates a view at low power.** (See Volume 3 Chapter 4.) Credit: Gerhard Bachmayer

Right: **In a large amateur telescope, the galaxy M51, the Whirlpool galaxy – a favorite for northern hemisphere observers – looks similar to this at medium power. The first time you see a bright galaxy is a memorable occasion. I will never forget as a teenager when I first saw the Pinwheel Galaxy which lies 3 million light years away in my small 100 mm (4") telescope. I was amazed that I could see the giant nebula that lies in its outer spiral arm.** (See Volume 3 Chapter 8.)

GETTING STARTED

PLANETARIUMS

It's a good idea for novices to visit a planetarium to learn about the night sky. Advanced amateurs and those in the general public also thoroughly enjoy the experience. By using special projectors that project stars onto a large dome, Planetariums create such a realistic simulation of the night sky that it gives you chills up your spine, especially when they show what the sky is like in a dark country location. The simulation comes complete with the Milky Way, the planets, shooting stars, clouds, the Moon, and the Sun. Planetariums can simulate light-polluted city skies and then demonstrate what a surprising difference it makes to be under a naturally dark, star-studded sky. The planetarium's director will point out the constellations, and many objects of interest, as well as showing the audience special events that occur in the heavens. You'll learn how the night sky is constantly moving and changing. Planetariums can even take you backward or forward in time! Large modern planetariums project breathtaking space edutainment movies onto their domes to take you on fantastic, Imax journeys through the solar system and to the far reaches of the universe. With the dome imagery occupying all your vision, you can feel like you are flying through space!

Many planetariums have an observatory with a large telescope for public viewing nights. Some also have outdoor observing nights where local astronomers show visitors objects in their telescopes and provide advice to novice stargazers.

The half meter (22) Zeiss Cassegrain telescope at the Stardome Planetarium in Auckland, New Zealand. Credit: Gregg Thompson

Planetariums are excellent venues to learn about astronomy. They create an artificial but very realistic view of the night sky. Many like this one in Brisbane, Australia have an observatory attached.

At the Milwaukee Public Museum, an Imax planetarium movie takes the audience on a stunning joyride through the universe. Credit: Evans and Sutherland

ASTRONOMY CLUBS

Local astronomy clubs often advertise public observing nights. At these, you get to talk to amateur astronomers who can teach you about observing with a telescope as well as buying and building them. Most clubs have dark sky observing excursions over long weekends. Some members take impressive astrophotography of deep sky objects. At their monthly meetings, experts give talks on a broad diversity of subjects. The larger clubs often have their own observatory that houses a large telescope for members to use. Astronomy clubs typically have a library from which you can borrow books, magazines, and videos. To find astronomy clubs worldwide, go online and search 'astronomy clubs/associations'.

GETTING ADVICE

If you subscribe to one or two **astronomy magazines**, you will learn a great deal about all aspects of astronomy. Many of the numerous free, online astronomy websites are also very good places to learn about individual subjects. You'll find that a good **star atlas** is a big help for finding your way around the sky. (See Chapter 5.) *STARGAZING* provides most of this information and much more.

It's a good idea to start stargazing by observing with your naked eye. Large objects like the Milky Way and large naked eye comets are best seen with the naked eye. A pair of binoculars are good for looking at faint naked eye objects to see them in more detail, and it's a treat to use them to wander through the star clouds of our galaxy to see so much more than the eye alone can see. Once you have had a good look around both the summer and winter skies in binoculars you may then want to start observing with a telescope that will allow you to see thousands of objects across the universe in considerable detail. (To know what you can see with the naked eye, binoculars, and telescopes, see Chapter 4.)

From a city location, you will be able to see the Moon and the naked-eye planets quite well because they are bright however, when you want to view faint objects such as comets and objects in deep space, it is best to observe them under a dark sky so their faint regions can contrast against a dark background. Deep sky filters will overcome moderate light pollution when observing nebulas in urban areas, but not star clusters or galaxies which require a dark sky to see them well. (See Chapter 9.)

SHARING KNOWLEDGE AND IDEAS

Stargazing can be truly awesome experience when you are alone under the stars. It can be even more rewarding when you share your sense of wonder and discovery with other like-minded people.

Stargazing can involve the excitement of planning a weekend **astro-camp** away with friends or a partner. It is also about getting together for a barbecue or watching a new documentary or movie on space, and then discussing aspects of it afterward. Stargazing is also about exchanging information with other astronomers across the planet via the Internet. It's rewarding to share and discuss the latest amazing space telescope images with others and those that advanced astrophotographers are taking. There is an explosion of amateur astronomers obsessed with the beauty they record in their astrophotographs.

Many amateurs love exchanging telescope making designs and construction techniques for building observatories, while others write computer programs for their observations. (See Chapter 11.)

When you go to astronomy club meetings, conferences, or to inter-club dark-sky star parties, you will have the opportunity of meeting many skillful amateur astronomers with whom you can exchange ideas and learn a lot from. Serious amateurs conduct valuable research, and some make discoveries. At the meetings of the larger clubs, you will get to meet and listen to lectures by high profile scientists who are at the cutting edge of astronomical research. From time to time, clubs have talks by visiting professional astronomers, authors of astronomical books, science media presenters, and sometimes astronauts.

Amateur astronomers at a Greek Star Party on Mount Parnon at 1,420 m (4,660') are enjoying dark skies. Amateur astronomers use dim red lights in order to maintain their night vision. Credit: Panagiotis Katsichtis.

Observers at the Texas Star Party take advantage of the dark skies to observe deep sky objects. At star parties, amateur astronomers share the experience of stargazing objects while also making new friends. Credit: Ron Ronhaar and Todd Hargis

You'll find that the social side of stargazing can be most entertaining, educational, and a lot of fun. You will make a lot of friends.

STARGAZING IS A MULTI-SENSORY EXPERIENCE

While there are many images of space in astronomy books, magazines, and on the Internet, there is no substitute for experiencing the night sky first-hand with your own eyes.

When you look at a photograph of the night sky, the image typically occupies less than 20° of your vision, but when you are out under the sky, it goes from horizon to horizon giving you a marvelous 160° visual experience! And the detail seen with your naked eye is far greater than what you can see in full sky images that are reproduced in books or on the Internet.

When you are out under the stars, you often see interesting dark silhouettes of trees, nearby buildings, or distant mountains contrasting against the soft glow of the Milky Way and thousands of twinkling stars. Foreground features create great 3D depth to the scene and they add much atmosphere that you do not get when looking at pictures in a book or on a screen. While many photographs are very atmospheric and beautiful, they cannot capture the feeling of being outdoors seeing the real universe. (See Chapter 10.)

When looking at a photograph, you only use your sense of sight, whereas when you are outdoors, you are using multiple senses. You hear the sounds of nature all around you. A bird call can travel for a mile at night. You smell fragrances from plants and trees or the aroma of

cattle in a nearby field, or perhaps the inviting smell of a campfire or barbecue. You feel the temperature of the air and it moving around you. Unfamiliarity with a new environment at night makes your senses more alert so you take in more information. The specialness of being in the dark under the stars is so much more romantic and atmospheric than looking at a photograph.

As the night progresses, you'll notice that the Earth's rotation is causing the canopy of the sky to slowly move westward, thereby bringing into view new constellations in the east. You see all the constellations as they really are in relation to one another in full surround vision. No picture can reproduce that.

When you look through a telescope to see many amazing aspects of the universe, you are one of a very small percentage of people on Earth that have had the opportunity to do that!

Viewing the sky live, and looking at excellent photographs, both have their advantages, so they are not exclusive. One can see things in photographs from space or long exposures that the human eye can never hope to see, but equally, you can experience aspects of astronomy when viewing the sky with your eyes that no photograph can ever capture. To maximize your enjoyment of astronomy, the two should go hand in hand.

Throughout the STARGAZING volumes, there are images of open sky scenes like the one below. These should encourage you to visit dark sky sites to see the beauty of the universe uncompromised by urban lights.

I hope the cosmic joyride you about to take through reading STARGAZING will excite you, and that the knowledge you gain will make your journey a rich and fulfilling one.

Experiencing dream-like scenes like this greatly enhances the joy of stargazing.

This dreamlike scene shows the Subaru Telescope on top of the 4,000m (14,000') high extinct volcano on Mauna Kea on the Big Island of Hawaii. Light from the town of Hilo at sea level illuminates the clouds from below. Above the clouds is a magnificent view of the center of our galaxy setting. Wally Pacholka's has many magnificent scenes like this one which are available as prints from Astropics.

Chapter 2
THE BASIC BUILDING BLOCKS OF THE UNIVERSE

To understand the structure of the universe, it helps to know its main components. Once you can visualize the seven main building °blocks below, it becomes easy to grasp how they form the universe that we see. Gravity is the force that brings together these building blocks.

This chapter describes the major categories of objects starting from the smallest objects such as stars to the largest, which is the structure of the universe itself. Objects smaller than stars such as planets, moons, asteroids, and comets, are included under Solar Systems.

To get a good mind-picture of the sizes of things in the universe, look at the many websites that show the relative sizes of planets, stars, and to get a good mind-picture of the sizes of things in the universe, look at the many websites that show the relative sizes of planets, stars, and galaxies by typing in '*size of the universe*'. There are impressive sites that allow you to compare all objects from the most colossal in the universe right down to the most infinitesimally small in the sub-atomic quantum universe! (See Volume 3 Chapter 1.) One very good and very popular video goes from planetary bodies in our solar system and it compares them to the sizes of many types of stars in our galaxy. From there it takes its audience on a journey to the edge of the universe comparing our galaxy to that of the entire observable universe. To watch this online, search for '*Star Size Comparison 2*'.

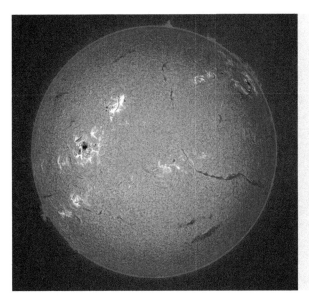

STARS

Our Sun is a medium-sized star. The largest super-giant stars have volumes billions of times larger than our Sun, while dwarf stars have volumes up to fifty thousand times smaller.

Some stars are far hotter and brighter than the Sun, although the bulk of them are cooler and dimmer. There are stars that are so bizarre that several decades ago, their properties would not have been thought possible. (For details about stars, see Volume 2 Chapter 3, Volume 3 Chapter 1, & Chapter 2.) Credit: NASA

SOLAR SYSTEMS

The term 'solar system' describes planetary bodies in orbit around one or more central stars. Solar systems condense from huge clouds of gas and dust called nebulas. As the gas contracts under gravity, the cloud begins to rotate. The density in the core becomes so great that nuclear fusion occurs and a star is born. A disc of matter orbits around the newly formed star. Planets, moons, asteroids, and comets condense in the disc. (See Volume 2 Chapter 6). It appears that almost all stars will form solar systems of planetary bodies. The diameter of a solar system typically extends from less than half a light year for the smallest stars, up to a few light years for the largest ones. Our solar system is around one light year across. (A light year is the distance light travels in one year.) Credit: NASA

NEBULAS – INTERSTELLAR CLOUDS OF GAS AND DUST

Solar systems condense from gas and dust clouds called nebulas. (See Volume 2 Chapter 6.) Inside them, many stars form in clusters, as seen here. (See Volume 3 Chapter 6.) Nebulas exist in the spiral arms of galaxies. They vary in size from ten to several hundred light years across. (See Volume Volume 3 Chapter 4.) Credit: HST, NASA

GALAXIES

Galaxies are collections of large populations of stars and nebulas, as seen here as pink regions. The smallest dwarf galaxies have a few billion stars while our Milky Way has around a trillion. The largest elliptical galaxies have up to a hundred trillion.

Irregular dwarf galaxies may be as small as 1,000 light years across whereas the spiral arms of our average-sized Milky Way galaxy are almost 100,000 light years in diameter. The largest known galaxy is estimated to be five million light years in diameter!

The closest major galaxies are a few million light years away, whereas the most distant ones are close to the edge of the observable universe at over 12 billion light years distant. (See Volume 3 Chapter 8.) Credit: Galaxy M33 by Robert Gendler

CLUSTERS OF GALAXIES

Over billions of years, gravitational attraction draws galaxies together to form a cluster in which dwarf galaxies orbit larger galaxies. Eventually all galaxies in a cluster will merge into one very large super-massive elliptical galaxy. (See Volume 3 Chapter 8.) Credit: Stephan's Quintet by HST NASA

SUPERCLUSTERS OF GALAXIES

Galaxy clusters also gravitate together to form superclusters. These can contain from two to twenty galaxy clusters. A supercluster of galaxies can contain tens of thousands of individual galaxies of all sizes. At the center of these superclusters are enormous elliptical galaxies that contain hundreds of trillions of stars. It is estimated that there may be around 10 million superclusters of galaxies in the part of the universe that we can observe known as the 'observable universe'. Astronomers are now discovering clusters of superclusters! (See Volume 4 Chapter 1.) Credit: Abell 2744 supercluster by HST, NASA

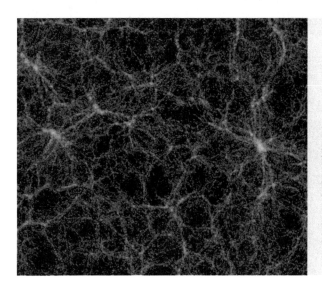

THE FILAMENTARY STRUCTURE OF GALAXY CLUSTERS

It's hard to imagine, but a seemingly endless number of galaxies fill our universe. They are not scattered randomly but rather, they form a cosmic web-like structure in all directions throughout the cosmos. (See Volume 4 Chapter 1.) This structure has been likened to bubble bath froth in which galaxies gravitate along the membranes of the bubbles and few exist inside the bubble voids.

EVERYTHING ORBITS SOMETHING

As you learn about objects in the universe, you will realize that **gravity causes everything in space to orbit something that is more massive**.

The following are examples:

- small asteroids orbit larger asteroids; asteroids orbit their star or a planet,
- moons orbit planets and planets orbit stars,
- small star systems often orbit larger star systems,
- millions to billions of stars orbit super-massive black holes at the centers of galaxies,
- small galaxies orbit larger galaxies,
- small groups of galaxies orbit large clusters of galaxies and clusters orbit supermassive clusters,
- superclusters of galaxies orbit supermassive superclusters.

EARTH'S LOCATION IN THE UNIVERSE

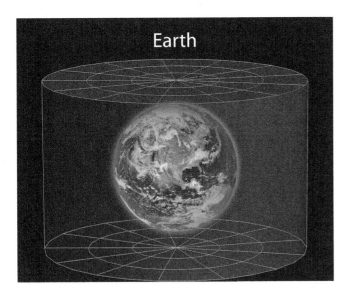
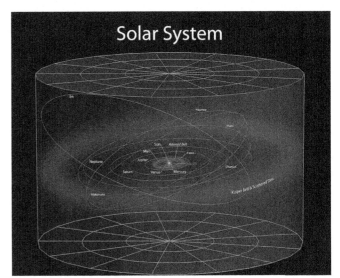
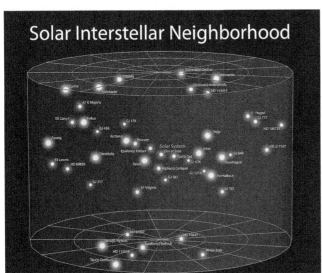

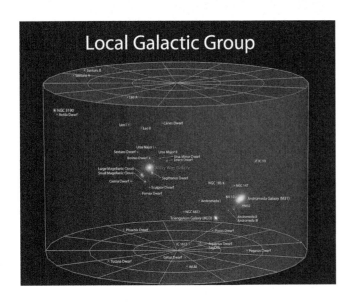
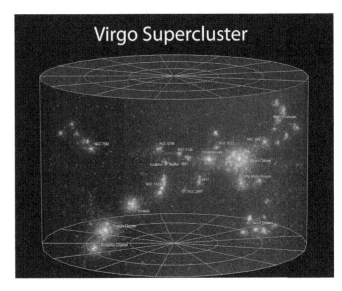

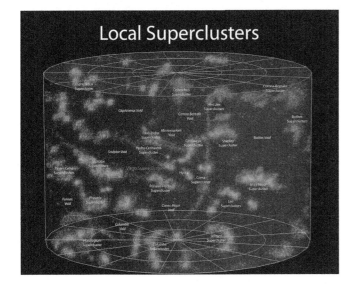
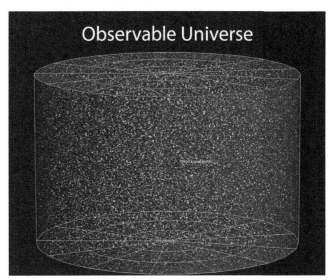

There is nothing in the universe that does not have a reason for the way it is. There is a reason for everything that happens, no matter how improbable that may seem to us. To know a reason for something, we just have to be smart enough to work it out. Of course, there will be reasons that are far beyond the intelligence and knowledge of all of humanity so we can never expect to know those reasons. The more we learn, the more complex everything in the universe becomes. The universe is not illogical or haphazard in the way it works. Scientific evidence is leading us to realize that it is an extremely fine-tuned program. But for what purpose?

"There is nothing that is not knowable if a mind has enough data and enough intelligence."
Gregg Thompson

Chapter 3
UNFORGETTABLE STARGAZING EXPERIENCES

Imagine being out in the desert under a cloudless, dark sky and you are staring into the galaxy while standing amongst surreal monumental geological features like those in Monument Valley. This would be so awesome that you would get shivers up your spine. In a location like this, you can imagine a star ship coming out of the heavens and landing out on the plain and you are the only one that knows about it.
Credit: Wally Pacholka /Astotpics

OBSERVING IN SURREAL LOCATIONS

MAKING THE MOST OF STARGAZING EXPERIENCES

Stargazers who experience extraordinary night scenes and special astronomical phenomena, often describe what they have witnessed with such words as *"unforgettable!" "just magical", "so romantic", "unreal"*, and sometimes, *'dream-like'* or *'ethereal'*. You too are likely to experience such feelings when you travel away from urban areas to stargaze in un-spoilt, natural open spaces where you are surrounded only by the sounds and sights of nature at night and the stars filling the heavens above. Despite the sensation of being immersed in what can be surreal environments, stargazers are nevertheless well aware that they are experiencing reality at its most wondrous.

To whet your appetite for enjoying extraordinary stargazing experiences, I recount here, and elsewhere in *STARGAZING*, just some of the many marvelous experiences I have had observing with stargazing buddies, my family, and with camping friends on special occasions. While the images accompanying the descriptions should help to convey how very special these experiences have been, they only provide a small impression of how all-encompassing such real-life experiences are. This is because, unlike seeing a photograph, real-world experiences engage all your senses in 360°, as well as your emotions, and your imagination. They can be truly awe-inspiring!

Stargazing is not simply about looking at the stars: it can also involve the excitement of being in a wide diversity of very unusual, and beautiful environments. Because astronomers like to observe in dark locations away from city lights, they also get to see many special things that few people are aware of, let alone have the thrill of experiencing. We will now look at some examples if this.

When you camp in the country under a crisp starry sky, it can feel like you are in another world, where dreams really do come true.

If you have strolled along a dreamlike, tropical beach at night under the light of a starry sky with your lover, this can feel like you are the only people in our universe. It can be especially romantic. Scenes like this can be reminiscent of the marvelous dreamy beach scene in the movie '*Contact*', as shown here.

You and your observing friends may have your telescopes set up on a farmland far from civilization when the Moon begins to rise over a distant horizon. You feel like you have stepped into a magical scene out of a sci-fi movie. The Moon lights the landscape with its silvery light in a way that makes you feel safe and at home on the world you love.

Perhaps you have driven up to one of the world's major observatories where the telescopes are larger than life. They are so high-tech that you feel as if you are a time traveler that has been transported into the future. Look carefully and you will see the size of the person to the telescope.

COMFORT

When you observe from a dark sky location, it's wise to plan ahead to maximize your experience. Take lay-back seating like a lightweight, folding sun-lounge, or an air mattress and pillow so you can lie flat and be relaxed when looking up into the heavens with your binoculars or your naked eye. Also have soft, dimmable red lighting to maintain your eye's adaption to faint light. If you play soft, tranquil, atmospheric background music, especially some with a spacey theme, this can really enhance your sense of awe, wonder, and grandeur as you look into our galaxy. Be sure to take your favorite beverages to drink and some snacks to nibble on, as this will help you stay fresh. It's best not to drink much alcohol as it is likely to make you sleepy and accident prone. Ensure you have additional warm clothing, as the temperature always drops as the night goes on. **Being comfortable is essential to making your experience much more enjoyable.** If you like to have thought-provoking chats under the stars, then there are some great subjects to discuss in Volume 4 Chapter 1.

To maximize dark-sky stargazing experiences, whenever possible, go to outstanding locations that in themselves create unforgettable, mood- enhancing atmospheres – places such as national parks, remote open fields, undeveloped beaches and islands, interesting deserts, or be on a boat away from light pollution. I have left out 'up into the mountains' as mountains have the habit of creating their own cloud. (See Chapter 6, page 96.) To have great experiences, it's best to be with relaxed people who are imaginative, and who have a curiosity about the grandeur of nature. They will be more likely to delight in having such unique experiences with you. Don't be with people who are oblivious to beauty and grandeur, as they will not appreciate the experience and they will spoil it for you.

To fully enjoy being under a beautiful night sky, especially when you are in a location that you have not been to before, allow yourself time to sink into the uniqueness of your setting. It can have a somewhat spiritual experience if you take in the atmosphere without the need to talk. Absorb the site's tranquility and enjoy the stillness as a special, rare experience in itself. Give yourself enough time to relax before doing telescopic observing so you are not in a rush, as rushing will break the spell that you are hopefully under.

As twilight becomes night, sit back and relax with a drink in hand as you watch the heavens slowly move across your window into the universe. You may be camped by a lake with a wide-open view of the Milky Way arching across the sky and reflecting in the lake's still water. Perhaps, there are dramatic mountains sleeping peacefully in the west with the stars slowly setting behind them. Lay back and look up into the zenith overhead to take in the wide expanse of the heavens for its sheer beauty. You'll notice satellites and the occasional meteor moving through the stars as the evening nightscape envelopes you. Experiences like this can be a very memorable, and rewarding because they are not available in suburbia.

Following is a small selection of just some of the many special experiences that I and others have had while stargazing. They may give you some ideas for the types of stargazing experiences you can enjoy.

TWO UNFORGETTABLE COMET HALLEY EVENTS

Comet Halley's 1986 return was the first time the comet had entered the inner solar system since its previous return 76 years earlier when the Earth went through its tail. On that occasion, some people bought gas masks naively believing they would be poisoned by its gases.

During the 1986 return, I had two unexpected experiences that I will always remember. The first was on the night of March 8. Some fellow observers and I were on a mountain in Southeast Queensland, Australia along with hundreds of international tourists hoping to see the most famous of all comets when Halley was at its best. My astronomer friends and I had dreamed of this since childhood.

To everyone's great disappointment, the mountaintop was covered in cloud mist. Knowing that mountains often attract their own cloud, even when their lower surroundings are cloud free, I decided that we should drive down to the base of the mountain in the hope of clear skies. To our great relief, the sky there was cloud free. Unfortunately, everyone on the mountain missed the amazing sky show that we were treated to.

Around 2 am, we found an open grassy paddock where we could lay down on a ground sheet with pillows under our heads to observe the spectacle of Halley's comet when it rose before morning twilight began. The location was ideal, so we had a box seat for the sight we were about to witness. While waiting to see Halley rise, we had interesting conversations about what other planets might be like, and whether people would ever go to Mars.

We observed Halley's Comet from a country field as a New Moon was rising. A couple of days either side of New Moon was when Halley was it its stunning best. Enlarge to see detail. Illustration by Gregg Thompson and Rebecca Gjerek

The scene before us consisted of trees that lined the banks of a stream winding through the far end of the field. A deciduous tree closer to us contrasted dramatically against the beauty of the Milky Way to the southeast. In the distance, there was a partially forested hill.

Two hours before the beginning of morning twilight, we noticed faint wisps of the comet's tail emanating vertically from behind a hill to the east. Our anticipation was now growing. As the Earth rotated eastward, the tail continued to rise, growing ever longer and brighter. It became more condensed as it tapered towards the comet's head, which was still well below the horizon. The length of the tail was impressive because it stretched nearly halfway from the horizon to the zenith overhead! It measured 40° - that's eighty times the size we see the Moon!!

From behind the hill to the right of Halley's tail, we noticed another soft glow growing steadily brighter. It had a curious dark, bluish-gray dome with two thin white horns on either side. It slowly rose over the hilltop. At first, we were surprised because it looked rather alien. But within seconds, we soon realized it was the dark side of a thin crescent New Moon rising. It was only a few degrees away from the comet's tail. What a sight! As the Moon rose higher, its dark side became plainly visible. It was illuminated by sunlight reflecting off the bright dayside of our Earth. This is known as Earthshine. (See Volume 2 Chapter 4, page 104). In our binoculars, the Moon looked unusually large in comparison to the trees on the hilltop that it dwarfed.

The best was yet to come because the comet's head was still below the horizon. Within a few minutes, the round haze around the comet's head, known as its coma, started to rise above the hilltop. Although the comet's head was on the horizon, it was far more impressive than we could ever have expected. Its size gave us goosebumps. To its faint outer edges, the coma was an amazing six times the diameter of the Moon! Comparing its size with the Moon made Halley's head look very large. (In the illustration, the Moon had to be enlarged to be able to see what it is.) At the center of Halley's coma lay its bright, star-like nucleus. In my large 20x80 binoculars, I could glimpse a small, fine jet of gas and dust being ejected from it. We were seeing its fan-shaped tail almost side-on, so this compressed it making it much more distinct than it would have been if we were seeing it face-on. At this time, the dust tail overlaid the gas tail. In a month's time, we would see the tail face-on and it would be much larger and fan-like, but much fainter. We were blessed with a truly stunning view of Comet Halley. It surpassed even our imaginative childhood dreams. (See Volume 2 Chapter 12.)

Unfortunately, most people missed seeing it during the couple of days before, during, and after New Moon when it was at its very best. In some cases, this was due to cloud or low-lying fog, but it was mainly caused by incorrect media reports saying that Halley would be best seen in the warmer early evening sky in the following two weeks when it was closer to Earth. As a result, most people didn't bother to get up in the cool, pre-dawn hours. This was a big mistake because over the next fortnight, the Moon became more illuminated each night making it much brighter as it moved towards being a Full Moon. It lit up the sky to the point that only the brightest stars were visible. The bright moonlit sky swamped the comet's tail making it impossible to see. And to make things worse, Halley's orbit took it across the brightest part of the Milky Way, which was as bright as Halley's tail. So, even though the Moon was rising later every night, Halley's huge tail now had to compete with an equally bright Milky Way. This made most of the tail invisible. During this time, only the very center of the comet's head was visible as a moderately bright hazy spot. To the naked eye, the head looked like the globular cluster Omega Centauri, or the open star cluster M7.

All comets need a dark sky to contrast against, so we had to wait for it to move away from the Milky Way. But by the time that happened, it was then moving well away from the Sun, so as large as its tail remained, it brightness faded quickly. As well, we were then seeing its tail face-on. The tail remained very long and wide for several days but it was faint requiring averted vision to see it well. (To learn about averted vision, see Chapter 7.)

Sadly, for those who waited to observe Halley after it moved into the evening sky, they missed this once-in-a-lifetime opportunity. Few of the them are likely to be alive when Halley returns in 2061-2.

My second experience with Halley was also most unexpected. Ten days after we saw Halley in the pre-dawn skies, my family and I were staying at a large hotel in Alice Springs – a small town in Central Australia. People from all over the world had come to 'The Alice' to go on to the Yulara resorts near Uluru (Ayers Rock) and the Olgas, some 300 km south of Alice Springs to observe Halley in telescopes under the guidance of myself and other amateur astronomers. I was engaged to organize this for a world travel company.

While at my hotel in Alice Springs, I ventured out onto the resort's golf course just before midnight to make my nightly observations of Halley with my large binoculars. Because it was now Full Moon, only the head of Halley was visible. In the semi-darkness, a voice cried out, "*If you're looking for the comet buddy, it hasn't come over yet. We've been out here all evening and haven't seen it.*" Looking around, I realized I was beside a crowd of around 70 international tourists, mainly Americans. They were sitting on and lying around one of the hotel's golf course greens and a sand bunker. I asked some of those close to me what they expected to see. They said, "*a fireball blazing across the sky like an F1-11 (fighter-bomber) with its after-burners on!*" I replied that the comet would not look like that. I showed them that it was up there in the sky almost overhead, and if they liked, I could show them where it was. This comment met with general disbelief until I explained that I was an astronomer and that I had been observing Halley every clear night. They seemed satisfied that I appeared to know what I was talking about. I invited the rest of the crowd to come closer, so I could point out where Halley was with my laser pointer and tell them about it.

When I showed them Halley's head, there was great disappointment. Under the brilliant Full Moon, all they could see with their naked eye was the center of the comet's head. Many felt like they had traveled across the world to see what appeared to be nothing more than a small, fuzzy ball of light. By sharing their binoculars around, they could see the comet's head much larger and brighter and some of the tail. This pleased them a little. I explained that the Full Moon had brightened the sky so much that it had overwhelmed the brightness of Halley's huge tail - in the same way that daylight outshines the stars. The Moon was also hiding the beautiful Milky Way, which was the same brightness as the comet. I let them know that in a few days' time when they get to Yulara, the Moon will be rising later each night. This will allow them to see Halley in a dark sky together with the Milky Way, and thousands of stars.

I described how large and relatively bright Halley was when I had observed it 2 weeks ago. My vivid descriptions of its appearance against the sky thrilled them. I explained that the diameter of the comet's head was over 60 times as large as our planet! And its tail extended for over a hundred million kilometers across the inner solar system! Most of them had no idea that Halley was that enormous. They were excited when I told them about the armada of space probes from different nations that were rendezvousing with Halley. They learned that the European Space Agency's Giotto probe was designed to fly through the dust and gas in the inner coma so it could go close to the comet's nucleus to photograph it. Mankind would then finally discover what lay at the very center of a comet. The nucleus is what produces the gas and dust that creates the comet's tail. (To learn about comets, go to Volume 2 Chapter 12.) They were quite excited to know this. Although we humans have observed comets since the beginning of humanity, this would be the first opportunity that we have had to discover what a comet's nucleus was like.

I gave my audience a mental image of what we expected the nucleus to appear like. I told them it would probably be cratered, and have mountains, and that it would consist of a conglomeration of pebbles, dust, and frozen gases. Geyser-like eruptions of gas and dust would shoot kilometers high from the comet's sunward-facing surface. (Indeed, this is what Giotto ended up photographing. Just as Giotto was approaching its closest encounter with Halley, a tiny pebble that was ejected from the nucleus hit Giotto at such high velocity that it threw it off course and into an uncontrollable spin. (For details, see Volume 2 Chapter 12, page 487.) I then told them how the comet would move across the sky over the following fortnight and I informed them that I and other astronomers would be showing tourists Halley in telescopes at the Yulara Comet Watch site where they were heading.

When these people gained an understanding of what Halley's Comet was like as well as its history, and what space scientist expected to discover when their spacecraft reached the comet, this completely changed their initial perception of it. The knowledge they gained gave them a powerful sense of discovery. I received warm expressions of sincere appreciation from them for turning their initial disappointment into what they said became a 'very special experience and

To experience the Leonid meteor shower, we lay on the ground marveling at the bursts of bright green meteors crossing the whole sky.
Illustration by Gregg Thompson, Rebecca Gjerek

learning opportunity'. This made me realize that it was what these people learned about the comet, not what they saw, that thrilled them so much. This experience taught me that **knowledge is everything when it comes to maximizing one's appreciation of the wonders of everything in nature.** This is especially true for all the chapters in Volumes 2 and 3 and especially Volume 4, in which the latest discoveries are stretching our belief systems beyond their normal limit.

THRILLED BY CELESTIAL FIREWORKS

When at their peak, the **Leonids meteor shower** is the most prolific of all showers. For a couple of nights every 33 years in mid-November, the Earth moves through the very center of a trail of dust particles left behind by **Comet Tempel-Tuttle's** passage through the inner solar system. As these particles enter our upper atmosphere, they burn up producing brilliant meteors commonly known as shooting or falling stars. (See Volume 2 Chapter 2, page 35.)

When the Leonids last peaked in 2001, the Earth went through the center of the trail for two nights. On the first night, I was amazed to see many bright meteors streaking right across the sky, even from the center of the brightly lit city of Brisbane! This inspired me to hopefully see them in their full glory from a dark country sky the following night. My wife and I organized a few friends who had never seen a meteor shower before to join me on a remote mountaintop for what turned out to be an amazing astronomical experience.

Following a wonderful meal at a restaurant on the edge of the mountain early that evening, we drove a few kilometers to a small country park on the mountain plateau that overlooked a beautiful river valley. Our site was perfect as it offered a wonderful open sky in all directions. We placed groundsheets on the grass and lay under a blanket with our heads on pillows. We positioned ourselves in a radiating star pattern so that our heads were close together, so we could talk easily, and to also record our observations on a tape recorder. Because each of us was looking in a different direction, between us we could see the whole sky, so we did not miss any meteors. Three other groups of people also arrived for the event. The sky was moonless, perfectly clear, and there were no clouds and no wind. We couldn't have asked for better conditions.

Around 11 pm, we observed the first batch of brilliant green meteors emanating from the constellation of Leo. They came out of the northern sky in bursts, sometimes with two or more of them shooting across the sky. To our surprise and delight, there was the occasional red one. Within half an hour, there were many bursts of meteors with the brightest ones crossing most of the sky! The very largest of them passed beyond the southern horizon. By 12.30 am, the radiant in Leo had risen higher in the sky so we could see meteors radiating straight down into the northern horizon. These were at the center of the radiant so they appeared almost star-like with no tails because they were pointing straight at us. They looked like stars brightening and fading quickly because they appeared motionless to us.

The meteors moved very fast. Whenever a few appeared at once, or there was a particularly bright one, claps and choruses of "*Oh Wow!*" went up from each group.

By 2 am, the shower had diminished to only an occasional meteor, so we retired for a well-earned night's sleep. We all thought that this was a very special and memorable after-dinner event that we would have to wait another 33 years to see again - that's assuming we will be alive and lucky enough to get such perfect observing conditions next time around.

CRUISING THROUGH THE GALAXY

I'll never forget the night that my friends and I felt like we were on an interstellar cruise. It was after a twilight barbecue on a luxury launch anchored in a sheltered bay on a remote island. As the sunset colors faded away at the end of twilight, the stars brightened, and the Milky Way above Orion glowed beautifully appearing like a star-sprinkled river of faint milky light across our celestial dome. The air was crystal clear, and the sky was cloudless and moonless. There was not the slightest breeze and there were no waves, so the ocean was so calm that it looked like a mirror! The stars reflected in it perfectly making us feel as if we were floating through the galaxy. We were surrounded by stars in every direction and the horizon was barely noticeable.

The ambiance that this scene created was so magical that we noticed that we were all whispering intuitively so that we would not risk breaking the spell we were under. We sat on the deck, delighting in our marvelous view of the galaxy's star clouds and its dust lanes. We saw a number of shooting stars while sipping our drinks and quietly chatting as I pointed out special features in the sky to my companions with laser pointer. We thoroughly enjoyed this experience until 11 pm when, to our surprise, the conditions abruptly changed to strong gale force winds and a dangerous swell! Those magical hours were over, but they would never be forgotten.

So we did not break the spell of feeling like we were floating through the heavens all around us, we were all whispering to one another.
Illustration by: Gregg Thompson, Rebecca Gjerek

Canoing on a still lake under a starry sky can be a truly awe-inspiring, and dream-like experience.

AN INTERSTELLAR CANOE RIDE

Another experience of being immersed in the universe occurred some years later when my son and I were canoing along a remote sinuous river lined with overhanging rainforest. We could hear the strange sounds of nocturnal nightlife emanating from the total darkness of the forest on either side. From time to time, a bat or an owl would swoop low overhead to check us out. We were far from the lights of civilization, so the stars were blazing brightly when we arrived at the end of the river where it entered into a large lake. When we paddled into this, we had as expansive view of the heavens above. The air was so still that it seemed unnatural. There was no wind, or waves, and no current at all. We stopped paddling, so we did not disturb the glass-like surface of the lake. We could then had a perfect reflection of the sky in the lake. The galaxy seemed to surround us in every direction! We could barely believe our eyes. We literally felt like we were floating through interstellar space in a canoe. The horizon was far away that there was almost no separation between the stars reflecting in the lake and those in the sky. It was just mesmerizing. When we slowly placed our paddle in the water to turn the canoe around to look behind us, the small waves created distorted space below us as if gravity waves were moving through the universe. My son wondered how he could ever describe this to his friends so they could really imagine it.

When we were motionless in the lake, we realized that more than 99% of people have probably never experienced this, but there would be sailors, hikers, and fishermen who must have witnessed this type of spellbinding illusion on nights as still as this.

From a headland overlooking the ocean, I once saw Jupiter and some bright stars reflecting in an exceptionally rare, perfectly calm ocean that looked more like oil than water.

DINING UNDER A HEAVENLY ORB

To allow some of my friends to experience the enchantment of a total eclipse of the Moon, my wife and I decided to have a lunar eclipse dinner party for six at a very special seaside location. Before sunset, we set up a table and chairs and a small portable barbecue on the water's edge of a white sandy beach near the mouth of a magnificent, small river. We arranged for our friends to join us before sunset for a candle-light dinner on a warm, calm evening. To our delight, the sky was perfectly clear. There was no wind, and the water was warm.

When our guests arrived, we enjoyed drinks and nibbles after watching the sun set over mountains in the west. After this, as twilight commenced, I directed my guests' attention to the east over the ocean to see the Earth's deep blue-gray shadow rising above the ocean with the pink and purple colors of the 'Belt of Venus' above it. (See Volume 2 Chapter 1.) Once twilight faded and the horizon over the ocean became dark, everyone started to notice something deep red starting to come out of the sea. It had a strange curved form. There were very distant broken clouds just above the horizon, but they were almost invisible to the naked eye due to night approaching. In my binoculars, one of the women thought this deep red, slightly domed object was something alien, and she became frightened. Unbeknown to my guests, clouds were obscuring the edges of this object rising out of the ocean. After taking turns at looking at it through my binoculars, my guests became disturbed because it was growing in size and changing its shape and color. It turned from deep red to orange and shortly afterward, the top became yellow. Some of them speculated that it might be a large fire on a ship over the horizon but the top edge was sharp and curved. It had a distinct mushroom-like shape at this point but, unbeknown to my friends, this was due to clouds obscuring its edges. One couple thought it might be the cloud of an atomic bomb rising beyond the horizon. One fellow was convinced it was a UFO. Not knowing what it was, they thought one of us should leave to report this unnerving phenomenon to the authorities. We did not have mobile phones at that time so I suggested that we should hold on to see what eventuates.

I had experienced similar reactions to this phenomenon on two other occasions before this. I did not let on that it was the Full Moon rising between clouds in order to see how long it would take them to realize what it was. The high density of the atmosphere immediately above the ocean made the rising tip of the Moon appear a dull, blood red color when it first appeared. The dense air flattened the Moon's round shape. Its underside was very flattened. The ever-changing broken clouds on the horizon cut into its edge on both sides continually changing its shape, so this really added to the confusion as to what it was. Understandably, it was not recognized to be the Moon. As it rose into less dense air, which was more transparent, it changed color becoming orange and then yellow on the top. As it rose higher and moved away from the clouds, its flattened roundish shape could be seen well. Its top had now become white and its Man-in-the-Moon feature could be seen.

It was then that my friends realized that it was the Full Moon rising. Knowing this made them laugh with a mixture of relief, surprise, and some embarrassment. They jibed me for not telling them what it was.

They wanted to know why it looked so different, so I told them the factors that led to their speculations. I explained that jumping to highly unlikely explanations is common when people lack familiarity with uncommon natural events, especially at night. They agreed that jumping to a UFO sighting as the explanation was all too easy to do. UFOs are an exciting explanation because it makes the observer feel special. And of course, they agreed that UFO observations are a good tale to recount!

THE APPEARANCE OF THE FULL MOON RISING OVER A PERFECT HORIZON

The images on the following page illustrate how the Full Moon changes in appearance as it rises over the ocean.

When observing the Moon rising through perfectly clear air at sea level, you will notice that it is noticeably flattened, especially across the bottom. This is caused by refraction due to the air being very much thicker when looking across the horizon at sea level. In this horizontal view, it is some 200 km (120 mi) thick.

1. When the Moon first appears over the ocean, it is a dull, blood red color, due to dense air close at the horizon and a high content of aerosols absorbing its light.

2. As it rises higher and half of it is above the horizon, the top half becomes orange. If parts of it are hidden by horizon cumulus clouds, it is not obvious that it is the Moon rising. It can look very strange.
3. Once it has cleared the horizon, it appears reddish at the bottom, then orange up to halfway and its top half appears yellow, while the very tip is likely to be white. You'll notice that it has a compressed shape, particularly at the bottom.
4. Once the Moon is its full diameter above the horizon, it becomes a brilliant silvery white in clean air. Its Man-in-the- Moon features now become quite obvious. Its light is now passing through air that is much less dense. The bottom may still be a little yellowish depending on the transparency of the air at the time.

If you have a lookout to the east where you can see the Full Moon rise over a clear, perfect horizon, it's fun to look up the time that it rises and have a few friends come to your site before the Moon starts to rise. Don't tell them this will happen. See what they think it is when it first appears. For those who have not seen this, it can look strange and startle them. To see the moon change color, you must have a perfect horizon as the color change happens within half a degree, or one Moon diameter. It's a good idea to have one or two pairs of binoculars and/or a small telescope to observe it.

These images illustrate how the Full Moon changes in appearance as it rises over the ocean. Illustrations by Gregg Thompson, Rebecca Gjerek

OUR MAIN COURSE WITH A BRONZE ORB SUSPENDED IN THE MILKY WAY

We decided to have our entrée as we watched the Moon's silvery light glinting off small waves towards the river's mouth. The moonlight on the waves in the water looked like a stairway to heaven.

For table lighting, we had wind-proof glass candle holders that provided good lighting to see our food. Because they had shades above the candle, this kept the light out of our eyes, so we could see the stars well.

For added sensation, we kicked off our shoes, so we could sit with our feet in the warm, shallow water to enjoy the pleasing sensation of it and the sand between our toes while we dined and laughed about the good times we've had.

Try to imagine the scene in front of us. There was the dramatic silhouette of tall Norfolk pines and a monolithic, pyramid-shaped rock formation near the river's mouth. In the middle of the river lay a white sandbar glowing softly in the moonlight. It contrasted against a dark, forested hillside in the background that was sprinkled with the interior lights of modern homes that were nestled amongst the trees. The women commented that they felt like they were in a romantic movie. It was quiet, except for the soft sound of distant night birds, the trickle of the water around our bare feet, and the soft, relaxing, background music we had for the occasion.

A total eclipse dinner party at a very romantic location. Illustrations by Gregg Thompson, Rebecca Gjerek

The best was yet to come! I had not told my guests that a total eclipse of the Moon was going to commence so that it would be a special surprise for them.

After we had finished our entrée, my guests noticed that the bright Full Moon had become partially shaded on one side. One of our guests was now concerned about what was happening to the Moon because she had never seen a Full Moon start to disappear! I explained that a total eclipse of the Moon was occurring due to the Moon entering the Earth's shadow. She was relieved and excited to hear this.

By the end of our main course, the Moon was wholly eclipsed. Half of it had become a beautiful copper-orange color with a soft white tip . The other half was almost black. This really impressed my guests, because all but one had never seen a total eclipse of the Moon before. They were fascinated when they understood what was occurring. By dessert, the Moon had entered the very center of the Earth's shadow making it a deep, dull red on one side, and on the other side closest to the very center of the shadow, it was so dark that it was invisible. Earlier in the night, the moonlit sky was very bright so only a few bright stars were visible, but now with the Moon being very dim, the sky was full of stars! The Milky Way was now visible due to there being almost no moonlight and, fortunately, there was little commercial development in the area, so there was no light pollution.

After dessert, the Moon began to brighten as it slowly returned to being full again. By the end of dessert, we were delighted with our evening's very special light show that nature arranged for us. My friends said it was the most memorable dining experience they had ever had, and one that they said they could never have anticipated.

Anyone who is interested in sharing a **total lunar eclipse** dinner experience with others should look up on the Internet when the next total lunar eclipse will occur in the early evening. Do not bother to do this for a partial eclipse as they are not that impressive. (See Volume 2 Chapter 4 for details on lunar eclipses). Of course, you will need a good clear sky and pleasant weather conditions.

Phases of a total eclipse of the Moon. Credit: P. Hart

STARGAZERS GET TO SEE NATURE'S NEONS

BIOLUMINESCENCE

Being in dark locations, stargazers not only get to see luminous nebulas in our galaxy, but sometimes they also get to see nature's biological light shows as well - that's provided that they are in regions where these life forms exist. They are typically seen in moist, temperate, tropical forests. If you know where to look, you will see bioluminescent life that most people have never seen.

If there is a rainforest near your observing site, then take a break when cloud comes over and go into the forest to look for bioluminescence.

An astronomer friend opened my eyes to this. He picked me up at night and said *"I'm taking you to see something magical"*. I had no idea what he was talking about. I said, *"But it's night!"* He said, *"Trust me"* so I did. My curiosity was very high. After a 40 minute drive, we reached the top of a heavily forested mountain. He stopped his car beside the entrance to a popular rainforest walk. He handed me a green Cyalume light stick and he had a blue one for himself. We proceeded to walk into a pitch-black forest. I asked him why we were not using flashlights. He said I would soon understand. He ushered me into to the darkness of the forest where the only light I could see was from my faint light stick reflecting off the surrounding plant growth and the stars peeping through gaps in the high forest canopy directly overhead. In such total darkness, the stars and the sky glow were more brilliant than I had ever seen them. He had bought a portable music player with very atmospheric music that he played softly as we walked though this dark wonderland. His selection of beautiful atmospheric synthetic music made the experience seem surreal. He said, *"So you maximize your night vision, hold your light stick so that it gives a soft light on the path, but do not let it shine in your eyes"*. A little way down the path, he pointed out a light on a tree trunk. I was excited because it was a large yellow-green bioluminescence fungi. It looked like two saucers stuck together with dark lines around its edge. I was surprised by how bright and colorful it was in total darkness. My friend had a small torch, so he showed me what it was like in white light and I immediately recognized it. Further on, we noticed the faint light in many parts of a dirt bank along the path. On close inspection, we could see that blue-white light was coming from hollows in the bank.

When we looked closely at these lights, we could see that each one was a glowworm suspended by a score of fine silk-like threads from the top of these small cave-like depression in the dirt bank. Hanging below the glowworm were numerous sticky, glass-like dew drops glistening along the threads from the light of the glowworm above.

In our magnifying glass, the dew drops looked like a grand chandelier! At the base of each hollow, there was what looked like a miniature fir tree forest. We could see that this 'forest' was actually a fern-like miniature moss.

On traveling further into the forest, we found clusters of bright turquoise mushrooms on dead logs. When seen up close, they looked amazing. He then told me to kneel and look at the dead leaves on the path. When I did, I saw that they were glowing faintly in green light around each part of the leaf that was being eaten away by millions of bioluminescent microbes. When we arrived at the creek, there was this mysterious greenish light swirling around in the rock pools. We went onto a cave near along the creek where a waterfall fell through a hole in a part of its ceiling. To our surprise, we felt like we were in some new form of outer space because there were innumerable glowworms covering the roof of this pitch black cave!

Years later, my wife and I wandered into a forest on a remote mountain at night where a sign said there were bioluminescent mushrooms and fungi. We were astounded by how many different types and colors there were. Apart from the green and turquoise ones I had seen elsewhere, there were others that were purplish-blue while some were yellow and orange. There were even glowing centipedes that had fluoro red dots and green heads! We were stunned to discover this place. Not even the locals knew about it because none of them had ever been game to venture into this forest at night.

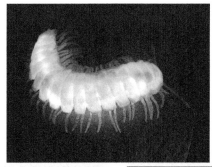
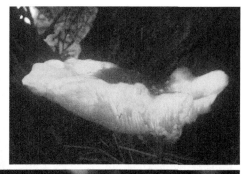

It is a wonderful visual diversion from stargazing to discover bioluminescent fungi, mushrooms, and centipedes in some forests near observing sites.

Bioluminescent life forms are amazing when viewed in the dark, and even more so when using a magnifying glass. Being in a bioluminescent forest is like being on another world. It will remind you of the fantastically conceived, bioluminescent forests in the movie '*Avatar*'.

When you walk through a forest at night that has bioluminescent light, it makes you feel like you could be on another planet. Credit: 20th Century Fox

FIREFLIES

In some places around the world in spring and summer, on a dark night, you may see fireflies flying around forest glades or across open grasslands. In some places, these glow-in-the-dark insects are not hard to find; they may even find you in your backyard. Near Austin Texas, I discovered a firefly called a 'lightning bug'. They swarm seasonally to search for mates by flashing their silvery white mini flashbulbs at the end of their tails. When I questioned my stargazer friends as to what I was seeing, and unbeknown to me, as a joke they made me believe that I was imagining these lights because they said they could not see them. While looking through my friend's telescope, a small insect landed on my hand and started flashing. I used a torch to look at it. My trickster friends laughed when they realized I had discovered what these curious flashing lights were.

Fireflies can swarm in some southern parts of North America as well as in Southeast Asia, Australia, Africa, and other regions that are typically away from highly developed areas.

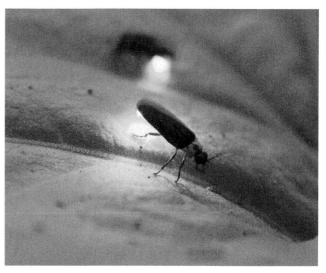

A firefly beetle has a bioluminescent tail that glows like a tiny Cyalume light stick. Credit: T Kennedy

In this short time-exposure, the yellow trails of fireflies can be seen flying around in a forest glen. In spring and summer, it's a marvelous experience to witness fireflies dancing across grassy fields or in forest glades. Credit: G Wald

Invasion from Space

One night I was asked by the weather bureau to investigate many UFO sightings in the sky. They were being phoned into the bureau from a beachside suburb about 60 km away. From my observatory, I could see nothing in the sky in that direction, so it could not be at a high altitude in the atmosphere, or out in space. It had to be a local phenomenon so I phoned an amateur astronomer friend who lived on a hilltop about 10 km away from where the reports were coming from. I asked him if he could see anything there in his telescope. He reported that he couldn't, so I then asked another amateur astronomer who lived only a couple of kilometers away to ride his bike over to where the reports were coming from and tell me what he could see. He phoned me from there to say that there were hundreds of people lying in their yards, on grassy footpaths, and on the beach watching a spectacle of thousands of yellowish-white star-like points of light moving around the sky amongst the stars! I was amazed. He informed me that these objects would stay still for short times and form geometrical patterns and then they would disperse to form different patterns. I had no idea how to explain this. I asked him if he could obtain a pair of binoculars from someone to see if could work out what they were. When he did, he told me that they looked like insects flying around only 20-30 meters or so up in the air. We later realized that it was a rare mating frenzy of fireflies. They were once plentiful in this region when it had been tropical rain forest several decades earlier before urban development had cleared much of the forest there away. However, there still are fireflies further away in the rainforest near the hinterland mountains.

Firefly swarms appear like numerous points of light flashing off and on, as they fly around in a forest. Their light is typically yellow but some are white and occasionally red ones are seen in the tropics. They are a wonderful diversion from stargazing when it is cloudy.
Credit: Topiat

The geometric patterns that people thought they were seeing were coincidental arrangements that were never perfectly geometric. This effect was due to wishful thinking and the fact that the human eye and brain are programmed to look for geometric patterns. This makes us see patterns when they don't really exist. There was considerable speculation by hippies in the area that the lights were UFOs in space that were getting ready for an invasion of Earth! It is common for some people to come up with extreme explanations in order to explain things which they have no knowledge about. In the years after this, I observed similar explosions of enormous numbers of other insects such as beetles and butterflies along the coast. They would die in their millions over one or two days and nights after their mating frenzy.

Glow-Worms

Glow-worms are fascinating to see. In forests, they can be found inhabiting cavities in dirt banks, in boulder grottos, old tree trunk hollows, and in large numbers in caves. They attach themselves to the ceiling of a cave or a cavity with their sticky silk. Glow-worm larvae hang silk-like threads from the end of their body. They have sticky droplets along them in order to catch insects that fly towards their light. When a flying insect becomes entangled in their threads, they pull the thread up and eat the insect - much like spiders do. When they mature, they turn into a flying insect so they can fly away to mate. Glow-worms can make a cave look like its full of rich star clusters.

Bioluminescent Plankton

From time to time, you might be observing under a dark sky that is beside the ocean or a lake. This might give you the opportunity to witness a bloom of rice-grain-sized plankton that emit a beautiful electric blue light when the water is disturbed.

When I visited the Great Barrier Reef in north-eastern Australia, some friends and I were on a launch anchored in a marvelous bay that had huge roundish boulders and a white sandy beach. It was a warm night so we decided to go swimming under the stars off the back of our boat. To our great surprise, as we dived into the water, it lit up as if it was electrified! Each movement we made created a dazzling display of bioluminescence from plankton that causes this effect. When we swam to the beach, we used them to paint glowing stripes on our bodies. In the dark, we appeared like alien creatures emerging from the sea. A holiday maker and his young son were fishing up at the end of the beach. When they saw our illuminated star-studded bodies moving around as we danced between one another while moving our limbs to and fro, it was impossible for them to tell what we were in the dark. We heard the father tell his son to throw down his rod and make a run for it to trees behind the beach. We could hear him shouting out in considerable fear to his son to "*run, run!*"

Many years later, on a dark beach lit only by starlight, a friend and I created a similar experience by splattering spots from a Cyalume light stick over us when we came out of the surf on a beach at a beautiful coastal National Park. In this case, there were two mature fishermen who became freaked out by what we looked like some alien life form crawling in and out of the edge of the ocean waves. We moved apart and up and down so it was impossible to tell from a distance that this 'creature' from the ocean was two people painted in bioluminescence for a light stick. The poor fishermen were very dismayed walking backward and then one of them said "*I'm getting out of here*" and

Glow-worms in a cave can look ethereal and magical. Credit: P Prue

he ran for the sand dunes. His mate immediately did the same running as fast as he could back to the campsite. The next morning, there was a lot of talk and irrational speculation about what the fishermen had witnessed. The alien monster they had witnessed had become many times bigger than us with huge tentacles. We listened with much interest trying to offer rational possible explanations, but no one was having any of that. Both these guys are probably still recalling the night they saw this huge, alien octopus-like creature come out of the ocean.

When there is no moonlight on a summer night at a remote beach away from light pollution and towns, it's not uncommon to see bioluminescent plankton making the waves glow as they break. When you walk through the shallows, it looks like you are wearing electric socks! There can be brilliant blooms of this plankton in saltwater lakes as well. (See Chapter 11, page 182.)

When I was stargazing on an island with other astronomer friends who had never seen this phenomenon, they didn't believe my explanation for what the blue stars scattered along the beach were. At first, they believed that I had sprinkled the beach the night before with some new luminous special effect to surprise them. When I showed them what the blue 'stars' were in my 40 times mini hand-held microscope, they were amazed. There are many other bioluminescent sea creatures like jellyfish and fish. They are occasionally seen in a calm sea at night when they come to the surface.

Bioluminescence in water is caused by plankton that emit blue light when the water is disturbed. Their magical light appears bright when seen under a dark night sky. Credit: P. Hart (left) M. Feinberg (right)

Left: **Bioluminescent shrimp left behind by a receding tide in Okayama in Japan appear as blue light mini waterfalls on rocks.** Credit: Kate Sierzputowski

Right: **Glowing plankton illuminate waves at night as they wash back and forth along the shores of Hong Kong.**

EARTH LIGHTS

There are places around the world where people sometimes witness an unpredictable form of electrified air scientifically known as **'earth lights'**. In parts of the US, they are called **'spook lights'** or **'ghost lights'**. In the UK and Ireland, they occur near swamps and mines where they are known as **'Will-o'-the-Wisp'**. In the remote open plains of Outback Australia, they are referred to by their aboriginal name, **Min Min lights**. Although rare, they have been observed in over 60 countries.

They appear as orbs of soft light about 300 mm in diameter floating about a meter or two above the ground and drifting along on the air flow. Their color can be a yellowish-green but in some places, they are bluish-white or an orange-red color that sometimes flickers.

They seem to occur after some slight movement of the Earth's crust occurs in regions where there are veins of quartz rock in the ground. It is thought that slight seismic movement creates a piezoelectric charge that electrifies pockets of air within cracks in the quartz rock. This 'bubble' of charged air is slightly warmer so it rises to the surface to appear as a round, glowing, fuzzy, orb-like 'bubble' that fades away at the edges. They usually last for about a minute until the charge dissipates, but some have been reported to last longer. Just like large detergent soap bubbles floating along on a zephyr, they tend to follow the terrain, moving with the airflow about a meter or two (3'- 6') above the ground.

If you are lucky enough to see this wonderful, silent, and harmless natural phenomenon, don't be frightened by it because it has nothing to do with UFOs or spirits, as many people have been led to believe.

A stargazing friend who is a commercial airline pilot told me about the time he experienced a similar phenomenon inside his commercial aircraft. It occurred at night

On rare occasions, at specific places under dark conditions, orbs of electrified air, sometimes referred to as 'earth lights', float across the landscape. Credit: Gregg Thompson

when he turned the cabin lights off so passengers could see a spectacular lightning display in a nearby storm cloud with the Milky Way above it. When his plane was unexpectedly hit by lightning, a bluish- green glowing sphere entered the passenger cabin via the metal doors on the front galley. The sphere floated slowly down the aisle in a swaying manner with passengers fearfully leaning away from it as it came towards them. When it finally touched the metal toilet door at the rear of the plane, it disappeared with a spark and the sound of a short circuit 'pop'. A black electrical discharge mark was left on both the galley and toilet doors. Other pilots have also witnessed this phenomenon.

AURORAS

Stargazers who observe from locations at high latitudes will see another form of nature's neons - spectacular auroras that color the sky with moving curtains of green light above which there can be white beams and broad sheets of red light tinged with purple at the top. (Auroras are described in detail in Volume 2 Chapter 1, page 11.)

Spectacular auroras like this one occur at the edge of space. This one displayed three colors, each one being at a different altitude.

SPECTACULAR LIGHTNING DISPLAYS

Those who have observing sites with low horizons are often rewarded with views of nature's most dramatic light shows in the form of lightning brewing in storm clouds. Large thunderheads are brilliantly illuminated with rapid cloud-to-cloud discharges as well as producing startling cloud-to-ground strikes. Sometimes lightning crawls over the outside of a cloud.

On rare occasions, you may see lightning discharging from a cloud into clear air. Florida in the USA and many places in the tropics and sub-tropics are prime locations to see intense electrical storms. They can look exciting to the naked eye but watching a distant active storm cloud in mounted binoculars can be a stunningly impressive diversion for stargazers while waiting for cloud to clear.

By far, one of the two most dramatic and truly awesome lightning display I have ever witnessed in my decades of observing storms was when I was about to begin observing galaxies one night. The weather bureau announced on the nightly news that there was a large chain of storms approaching. They reported that the cloud tops were extraordinarily high reaching 18,000 m (60,000'). Typically, they are 13,000 m (40,000') high. Luckily, my observatory had a near perfect horizon in all directions from which to watch this storm. I first saw the storm cells approaching from the north. They lit up very distant cloud tops over mountains that were around 90 km (55 mi) away. Despite the air being very clear, impurities in the atmosphere made the very distant lightning in this storm cell distinctly red, which is rare to see.

As I watched a nearly continuous display of lightning, I saw one storm cell ignite a nearer adjoining cell. The lightning had now become brighter and its color was now pink. There were fingers of lightning running along the underside of this storm cell's high anvil. To my amazement and shock, a cell overhead exploded with brilliant blue spider lightning that ran across much of the sky! This all-encompassing display was amazing. It was accompanied by the delayed thunder which cracked and hissed, something that is also rare. In short succession, there was another all-sky display and then another. These massive, all-consuming displays were very threatening - even to this seasoned lightning observer! The overhead lightning soon ignited cells immediately to the south,

Stargazers sometimes get to observe distant displays of lightning that are spectacular when viewed in binoculars. Credit: Gregg Thompson, Nicole Brooke

and then soon after the next one further south. The color of the lightning reversed back to yellow and then red as it moved to the southern horizon. There were no ground strikes or rain throughout this whole display. This was surely a once-in-a-lifetime lightning experience. However, only recently, I witnessed another incredible display of spider lightning that kept repeating over much of sky for twenty minutes before it moved northward. It produced very loud thunder as it ran along the underside of the cloud base which was only 10,000' or less above us. Surprisingly, there were very few ground strikes. People were outside everywhere watching this spectacle and yelling out in total shock as each one occurred. It made front page news the next day. On YouTube, you can find many examples of videos taken by astronauts in orbit that show lightning storms igniting one storm cell after another. It's quite another thing though to see this from directly underneath the biggest of these storms! (For spider lightning, see page 46.)

In Venezuela where the Catatumbo River runs into Lake Maracaibo, there is an extraordinary number of lightning storms where there are thousands of lightning strikes illuminating the sky there for nine hours a night. This location has the highest concentration of lightning on the planet. For those who like to photograph lightning, this is the place to go. The storms ease off in the dry months of January and February.

Stargazers at dark sky sites can take impressive photographs of lightning displays. Credit: National Geographic (left), Gregg Thompson (right)

Here we see a chain of storm cells seen from space. Lighting flows from one storm cell to the next. Credit: NASA

Fireball lightning is very rare and short-lived. What conditions exist to causes it to occur is unknown.

This time exposure taken over 15 minutes from a city hill top lookout shows the progression of lightning strikes across this city. Credit: Vasin Lee

Rare spider lightning running along the underside of high cloud can cover so much of the sky that it makes you feel like it is going to consume you.

A pink or blue coronal glow erupts after lightning strikes a power grid transformer. Credit: Gregg Thompson, Nicole Brooke

Stargazers often see spectacular displays of lightning like this over Port Phillip Bay in Melbourne, Australia. Credit: James Collier

The thunderous explosions from the many ground strikes in this spectacular lightning storm early in the morning woke up many of the city's residents. Credit: Ann Van Breemen

WARNING: Distant lightning storms are safe to observe but those that have lightning strikes within 5 km (3 mi) can move closer to you quickly and produce dangerous ground strikes. If you are out in an open space with nothing high close to you that can attract a lightning bolt, a charge in the ground could form **a skyward 'leader' arc** that travels through your body and towards the sky. This is because you or your telescope, could be the highest point around, so don't be out in the open observing when a thunderstorm is close by. Leader arcs are dull and nowhere near as brilliant and powerful as a downward strike from a cloud, but leaders can carry hundreds of amps that can kill you. Leaders and side flash/splash strikes are responsible for 60% to 80% of deaths and

A leader stroke

injuries from lightning - not direct hits from the main strike. Storm watchers have recorded videos of leaders leaving objects just above the ground on a flat area.

There are also side flashes, known as splash strikes. These come off a main strike. These are also deadly. When a lightning strike hits the ground, it saturates it in 20,000 or more amperes of current. This spreads across the ground decreasing with distance, but it this charge that also kills people. While an electrical storm is passing over, sit safely in a car or inside a building.

The metal roof and sides of a car are what protect you from lightning, not the tires, as was once thought to be the case. The dominant attractors for lightning strikes are a high location, a pointed shape, and metal towers, fences, and railings that are grounded. During a lightning storm, stay away from plumbing, and do not shower or wash your hands. Keep away from windows and doors that are not sealed where lightning could enter. Do not lie near concrete walls as they can conduct lightning. Bring pets inside. Before a lightning storm hits, unplug electronic equipment. Metal objects on you such as jewelry or a phone do not attract lightning. Stay away from corded phones. Smartphones are OK. Lightning can strike the same place often, particularly if it is high.

% Lightning Deaths, Injuries, and Causes

Direct lightning strike	3-5%
Ground current from a strike	50-55%
Side splash lightning	30-35%
Upward leader lightning	10-15%

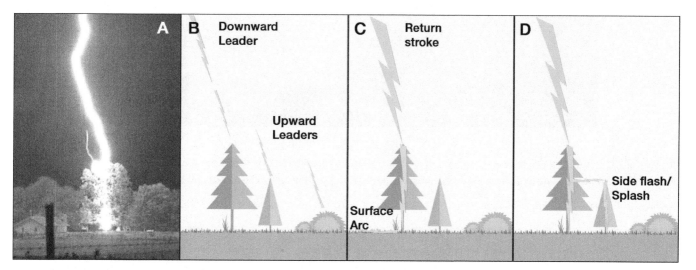

A. Lightning striking a tall tree. There is a small, faint leader from the tree that did not connect to the main downward leader.

B. Some upward leaders do not attach to the downward leader, but they still carry hundreds of amperes, so they are quite dangerous.

C. A surface arc is associated with ground current and can go tens of meters from the strike point.

D. Side flashes splash some of the current onto a nearby object as an additional path to the ground.

The average lightning bolt is about 2.5 cm wide and 8 km long

The longest bolt ever recorded crossed 190 km (118 mi)

2.5 cm (1 in)

8 km (5 mi)

Bottom Left: **A lightning strike hitting this grass left a large spreading burn mark. This shows the spread of the lightning dissipating in the ground. Anyone standing in his area would be injured or killed. They do not have to be directly hit by the lightning strike to be killed.**

Bottom Right: **This still is from an amazing video of a storm that shows rare, intense ground to cloud lightning.**

Coronal Glows

When viewing an electrical storm passing through a city from a high vantage point, I have often seen lightning striking kilometers in the distance. When it hits power lines and jumps insulators, it causes a bright, intense, hemispherical coronal glow for a couple of seconds before it quickly collapses. I have not discovered the reason why these events are blue and at other times pink.

Fireballs

When setting up to observe at sunset at a beachside campsite, myself and many others, witnessed a brilliant red fireball sent earthward from a violent storm cell only a kilometer away. The fireball hit the ocean and bounced a couple of times close to where lifesavers were training in their surf boat. It lit up the entire area for a couple of kilometers around in a brilliant orange-red light that startled everyone, even those inside homes and those camping in caravans and tents. It frightened the lifesavers so much that to onlookers, they appeared to row their boat back to the beach so fast that many of those who watched them joked about how they rowed it across the sand and up into the clubhouse! During an intense electrical storm in Sydney, astronomer friends reported that they were driving through a suburb at night in light rain when a bright blue fireball hit the wet street in front of their car. It bounced along for a few seconds and disappeared with a loud bang.

St Elmo's Fire

During intense electrical storms, three times I have seen city water towers on hilltops glow with a fuzzy, silvery-green light, which is then discharged skyward somewhat slowly in a zigzag form. This phenomenon may be a form of St Elmo's fire. On one occasion during an intense electrical storm, an astronomer friend was in his car at the beginning of nightfall when his car was hit by lightning. I was nearby sheltering behind a brick wall when the lightning hit. I saw my friend's car glowing blue with St Elmo's Fire for several seconds. I thought he must have been electrocuted but his car insulated him, so he was unhurt - except for the ringing in his eardrums, which I also had! No one has been successful in photographing this elusive fuzzy glow. Photos that claim to be of St Elmo's fire are of common lighting.

Ships, planes, cars, towers, etc. that have received a high static charge of electricity from nearby lightning during a storm, will glow off a blue-green light. This is known as St Elmo's Fire. Oxygen in the air touching such charged surfaces will fluoresce for a short time until the charge dissipates. Illustration by allswalls.com, Gregg Thompson, Nicole Brooke

UNFORGETTABLE EXPERIENCES VISITING MAJOR OBSERVATORIES

Stargazers that visit the world's largest observatories can also have unforgettable, dream-like experiences when they are there at night. To be at high altitudes, often in exceptional natural surroundings, and seeing huge, state-of-the-art telescopes is something few people get to witness. It is not easy to have the opportunity to be able to see these sites at night, but if you can get permission do this under a dark, clear sky, then you are likely to feel that you are at a truly surreal location. Such unusual, and often futuristic environments, enhance your imagination. They give you a sense of experiencing something very grand. Still photography and time-lapse video taken from these sites can produce stunning results.

When a friend and I were at the world-famous **Keck Observatory** on the summit of **Mauna Kea** at sunset, the scene looked very much like the image following. When the sky was becoming dark, we stood close to one of two huge Keck telescope observatories. They looked enormous against the heavens above. Work lights were on inside the dome while the astronomers and night assistants were getting ready for the night's first observing run. With the dome open, we could see the telescope's monstrous frames moving to where they would start their first observation. The work lights went out and numerous faint stars had now become visible in the now dark sky. A surprisingly bright display of the Zodiacal Light graced the western sky. (See Volume 2 Chapter 4.) The silhouettes of the huge black observatory domes contrasted dramatically against the brilliance of Milky Way. In the faint light from the night sky, we looked at one another thinking how far astronomy had come since we were teenage astronomers looking at the Moon with

A colorful sunset over Mauna Kea Observatory on the Big Island of Hawaii

our small telescopes. I have had similar, yet different experiences when observing at a number of other major observatories. (For details on major observatories, see Chapter 13.)

I could fill an entire book recounting many other great experiences I have had stargazing like those in this chapter and others described elsewhere in other chapters. Such experiences can be well described as being awe-inspiring, thought-provoking, unexpected, sometimes eerie, often romantic, and always exciting. Be sure to have your own similar experiences when stargazing. They will be unforgettable!

The famous, quintessential 200" Mt Palomar telescope observatory was the largest telescope in the world for decades. After the 1940's, Mt Palomar observatory symbolized the essence of astronomical discovery. It created the image of the archetypal mountaintop observatory where the huge 200" telescope would peer deep into the universe. It became a romantic vision that inspired many amateur astronomers and astronomy clubs to build their own domed observatories in its image. Credit: Wally Pacholka

Chapter 4
WHAT WONDERS CAN WE SEE IN THE SKY?

The Perseids meteor shower Credit: E Ivanov

A romantic view of the Milky Way over farmland. Credit: Miguel Claro

There is a great variety of objects and special phenomena to observe in the sky. In this chapter, they are listed in three groups: 1. those that you can see with your naked eye, 2. those best seen in binoculars, and 3. those seen in a telescope. The chapters in Volume 2 describe in detail the numerous objects and phenomena that can observe in our solar system, while Volume 3 covers what we can see in our Milky Way and in intergalactic space.

WHAT THE NAKED EYE CAN SEE

With just your naked eye, you will be amazed at how much you can see in space when you know what to look for. Following is a list of what is visible with the naked eye.

1. **The Milky Way Galaxy:** Billions of stars create this milky river of soft light across the heavens. You will see dark sinuous dust lanes throughout the Milky Way. This is what stars and planets are formed from. There are hazy bright spots in the Milky Way. These that are gaseous nebulas, galactic star clusters, and globular star clusters. (See Volume 3 Chapter 4, Chapter 6 & Chapter 7.)
2. **The Magellanic Clouds:** These are two dwarf galaxies in the far southern sky that orbit our much larger Milky Way galaxy. (See Volume 3 Chapter 8, page 301)
3. **The Andromeda Galaxy:** This is a companion spiral galaxy to ours, which is considerably larger. It is 2.5 million light years away . See Volume 3 Chapter 8, page 303.)
4. **The Constellations:** The constellations that are easy to identify are those that have bright stars such as, Ursa Major Big (the Big Dipper), Cassiopeia, Orion, Canis Major, Scorpius, Sagittarius, Virgo, Bootes, Vulpecula, Auriga, Corvus, Delphini, and Crux (the Southern Cross). There are many more, but they take time to identify them. (See Volume 3 Chapter 3, page 136.)
5. **Colored stars:** Red supergiant stars like Antares, Betelgeuse and Regulus appear rusty orange. Blue giants like Spica and Sirius appear bluish-white. Dwarf stars like Alpha Centauri, Capella, and Procyon are yellow. (See Volume 3 Chapter 1, page 13.)
6. **The Moon's large impact basins:** These are seen as seas of lava that long ago solidified. They are commonly known as the 'Man in the Moon'. Two large ray craters can be glimpsed with the naked eye. (See Volume 2 Chapter 5, page 119.)
7. **The changing phases of the Moon:** The Moon's changing phases throughout the lunar month as the line of sunrise or sunset (the terminator) moves across the Moon's face. (See Volume 2 Chapter 4, page 102.)
8. **The naked eye planets Mercury, Venus, Mars, Jupiter, and Saturn:** Each has its own subtle color. As each planet moves along its orbit, you will see

it move through the background stars. The planets vary in brightness as they, and the Earth, move around the solar system changing their distances from one another. From time to time, they can come close together forming impressive conjunctions in the sky. (See Volume 2 Chapter 6 – Chapter 11.)

9. **Bright naked eye comets:** These are an impressive sight in a dark sky. They move through the stars from night to night. Most naked eye comets have tails. Because the largest and brightest extend across an expanse of sky, they are best seen with the naked eye. (See Volume 2 Chapter 12, page 481 onwards.)

10. **The Earth's Shadow and the Belt of Venus:** This phenomenon is visible in the atmosphere during twilight for around 20 minutes, after sunset, or before sunrise. (See Volume 2 Chapter 1, page 9.)

11. **Auroras:** These are colorful displays of light in the upper atmosphere typically seen at high latitudes. (See Volume 2 Chapter 1, page 11.)

12. **Airglow:** This appears as a dull light emitted by the upper atmosphere near the edge of space. It covers the whole sky and stops it from becoming jet black. It is due to electrically charged particles from the Sun making the upper atmosphere fluoresce. (Chapter 5, page 78.)

13. **The Zodiacal light and the Gegenschein:** This faint band of light extends upward from the where the Sun sets, or rises, and it goes right across the sky. It can only be seen in a very dark sky. It is caused by interplanetary dust and solar wind particles that the Sun illuminates. (See Volume 2 Chapter 4, page 108.)

14. **Meteor showers:** These occur when the Earth passes through a stream of debris left behind by comets that cross the Earth's orbit. (See Volume 2 Chapter 2, page 36.)

15. **Brilliant bolides:** These are meteors the size of small rocks interplanetary space that burns up when they hit the Earth's atmosphere. (See Volume 2 Chapter 2, page 33.)

16. **Magnetic solar storms:** These storms on the Sun are many times the size of the Earth. They appear as black spots on the Sun's face when it is in its active stage. They can be viewed through a Sun filter. They move across the Sun as it rotates. (See Volume 2 Chapter 3, page 68.)

17. **Solar eclipses:** These are caused when the Moon passes in front of the Sun. When a solar eclipse is total, they are spectacular. At this time, hot pink prominences can be seen extending from the Sun's limb and the Sun's faint blue outer atmosphere extends along its magnetic field. (See Volume 2 Chapter 3, page 89.)

18. **Lunar eclipses:** These occur when the Moon moves into the Earth's shadow. When there is a total lunar eclipse, the Moon becomes a deep orange color. (See Volume 4 Chapter 4, page 106.)

19. **Solar and lunar haloes:** These are rings of light around the Sun or the Moon caused by sunlight or moonlight passing through ice crystals in clouds in Earth's stratosphere. (See Volume 2 Chapter 1, page 21.)

20. **Satellites:** These orbit the Earth. The brightest are as bright or brighter than the brightest planets. When their solar panels reflect sunlight directly at you, they brighten greatly for a few seconds. Sometimes they create brilliant fireballs when they re-enter the atmosphere. (See Volume 2 Chapter 1, page 22.)

21. **The largest open star clusters:** The Seven Sisters and the Hyades star clusters are close enough that several of their brightest stars can be seen with the naked eye. The brightest of more distant star clusters appear to the naked eye as hazy spots in the Milky Way. (See Volume 3 Chapter 6.)

22. **The largest globular star clusters:** The brightest of these appear as faint fuzzy stars. (See Volume 3 Chapter 7.)

23. **Nebulas:** Large nebulas like Eta Carina, or M42 in the Belt of Orion, appear as small hazy stars to the naked eye. There are a number of others that are more distant that can be seen with the naked eye as bright spots throughout the Milky Way. (See Volume 3 Chapter 4.)

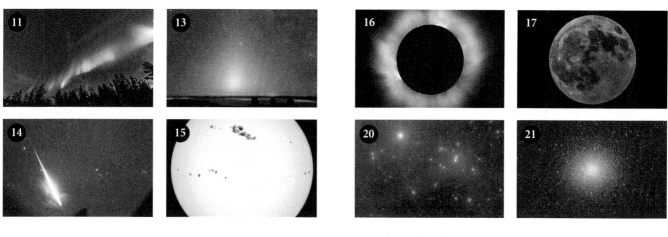

OBSERVING WITH BINOCULARS

The objects listed below can be seen in binoculars. Because binoculars have a lower magnification than telescopes, large objects are often best seen in them.

1. **The Moon's largest craters:** These are dark flat plains of ancient seas of lava seen as the Man-in-the-Moon. Several craters with explosive ray appear white in binoculars. (See Volume 2 Chapter 5.)
2. **Sunspots:** These can be seen in some detail when projected onto white paper or viewed directly with a solar filter over the binocular lenses. (See Volume 2 Chapter 3, page 72.)
3. **Large star clusters:** Dozens of the brighter members of the Seven Sisters and the Hyades are well resolved in binoculars. These clusters are so large that they extend beyond the field of view of most binoculars. There are many smaller, bright star clusters along the Milky Way that are well seen in binoculars as bright hazy patches. The brightest stars can be resolved in the closest ones. (See Volume 3 Chapter 6.)
4. **Gaseous nebulas:** The largest nebulas show some faint structure while most of the best nebulas appear as fairly easy-to-see bright, misty patches of light. (See Volume 3 Chapter 4.)
5. **Globular star clusters:** The brightest appear like fuzzy balls of soft light. (See Volume 3 Chapter 7.)
6. **The Milky Way's dust lanes:** Some of these show surprising detail in binoculars. (See Volume 3 Chapter 3.)
7. **Bright meteor trails:** Long lasting ones from very bright meteors show twisted contrails. (See Volume 2 Chapter 2, page 31.)

8. **Total eclipses of the Moon and Sun:** Binoculars show much more detail in solar prominences than with the naked eye. A totally eclipsed Moon is seen best through binoculars. (See Volume 2 Chapter 4 for lunar eclipses and Volume 2 Chapter 3 for solar eclipses.)
9. **Bright comets:** These are often best seen through binoculars because they extend over a few degrees of sky.
10. **The Galilean Moons of Jupiter:** They can be observed changing position hour by hour and day by day. (See Volume 2 Chapter 9, page 322.)

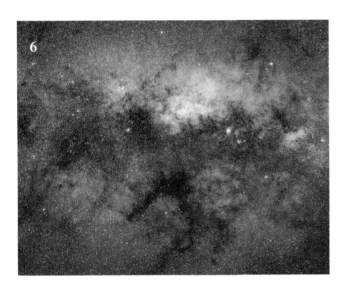

WHAT TELESCOPES REVEAL

Even a good small telescope will surprise you by to how much more detail and brightness it delivers compared to observing with binoculars. These are sights to observe in a telescope.

1. **Ever-changing views of storms in Jupiter's gas belts and the different appearance of its four largest moons that orbit it.** (See Voulme 2 Chapter 9, page 309.)
2. **Saturn's rings and moons:** Saturn presents an ever-changing viewing angle. Storms can sometimes be seen erupting and surrounding an entire gas belt. (Volume 2 Chapter 10, page 359.)
3. **Mars' markings, its ice caps and its dust storms:** Mars' changing seasons and its rotation are also clearly visible. (See Volume 2 Chapter 8.)
4. **Venus' and Mercury's phases:** These change as do their size due to the angle from which we see them as they orbit the Sun and so dos their sizes change considerably. (See Volume 2 Chapter 7.)
5. **Uranus and Neptune's discs:** These can be seen as small distant orbs. (See Volume 2 Chapter 11.)
6. **Pluto:** It appears as a very faint, star-like object. (See Volume 2 Chapter 12, page 439 onwards.)
7. **Asteroids:** These are visible as slowly moving star-like objects when they pass close to the Earth. (See Volume 2 Chapter 12, page 412 onwards.)
8. **The gas and dust tails of comets:** Jets can sometimes be detected emanating from the nucleus. (See Volume 2 Chapter 12, page 454 onwards.)
9. **Solar phenomena:** Much detail is visible in dark, complex magnetic storms known as sunspots. Brilliant white solar flares are sometimes visible. There are also huge flames known as prominences that erupt from the Sun's surface. Stunning detail can be seen with a solar telescope. (See Volume 2 Chapter 3.)
10. **Colors of double and multiple stars.** (See Volume 3 Chapter 1.)
11. **Globular star clusters:** Up to thousands of their brightest stars can be resolved in amateur telescopes. Hundreds of thousands of unresolved stars create a glowing orb of background light. (See Volume 3 Chapter 7.)
12. **Open star clusters:** Scores of moderately bright stars can be seen in these clusters. A number of the brightest stars are often orange or blue giants. (See Volume 3 Chapter 6.)
13. **Diffuse nebulas:** These are huge regions of bright and dark interstellar gas and dust clouds from which stars condense. (See Volume 3 Chapter 4.)
14. **Planetary nebulas:** These are gas envelopes ejected from giant stars. They display a marvelous diversity of forms. (See Volume 3 Chapter 5.)
15. **Novas:** Stars that infrequently have massive explosions across their surfaces causing them to become thousands of times brighter. (See Volume 3 Chapter 1.)
16. **Supernovas:** These stars are typically observed exploding in other galaxies. They become a million times brighter when they go supernova. (See Volume 3 Chapter 2, page 50.)

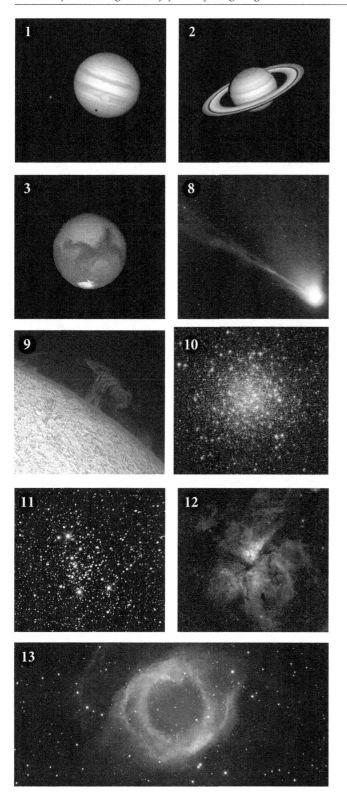

Readers who become very keen on seeing how far they can push their powers of observation, might like to read Phil Harrington's *Cosmic Challenge: The Ultimate Observing List for Amateurs*. Each chapter is segmented into all sorts of objects that are a test for categories based on aperture ranging from the naked eye to binoculars, to small telescopes right through to very large amateur telescopes. Finder charts and neat, accurate drawings of each object are provided. The book includes objects in the solar system, in our galaxy and right out to the most distant clusters of galaxies.

Now that you know what you can observe with our naked eye, and in binoculars, and telescopes, let's learn about how to go about observing to maximize what we can see in Chapter 6, Chapter 7, and Chapter 10. In Volume 2, you will learn about observing objects and phenomena in the Solar System while Volume 3 describes what you can see in deep space.

The center of the Milky Way reveals many star clusters and nebulas as well as dust lanes. Credit: Gerald Rheman

Christian Viladrich with a 1 m telescope

Chapter 5
ASTRONOMICAL NUMBERS, A LIGHT YEAR, STAR MAPS & ASTRONOMICAL TERMS

Part 1

ASTRONOMICAL NUMBERS

This part focuses on how to visualize numbers into the billions and trillions, as well as what a light year is, and how fast light travels.

When measuring extremely large distances across the universe and the enormous sizes of many objects, as well as the incredible energy outputs of highly energetic stars, astronomers have to deal with fantastically large numbers. They also have to think in timeframes that are often extremely long. These numbers are often in the millions, billions, and trillions, so it is very difficult for us humans to appreciate just how truly large such numbers are, particularly as we do not deal with such large numbers in our everyday lives. In astronomy, it's important to know what these numbers represent. Even though these numbers are frequently used nowadays when talking about economics and the debt levels of many countries and also large banks, it is surprising that most people, including too many politicians, do not know how large a billion is, let alone what a *trillion* represents. Even many mathematicians and astrophysicists who frequently deal with these numbers in their calculations, do not have any true appreciation of what these numbers represent in the real world. So, let's now look at just how large these numbers are.

A billion is a thousand lots of a million. That's a really large number, but a **trillion is a thousand lots of a billion** – now that's an extremely huge number. It's a number so large that it's difficult for most people, even astronomers to conceptualize its true size. The next level up is **a quadrillion. That's a thousand trillion – a number so large no human can visualize it!**

The following illustrations using money should help you to visualize these numbers up to a trillion. But to do so, we have used US$100 bills instead of $1 bills in order to fit these huge numbers on a page. **Due to $100 bills being used instead of $1 bills, try to remember that the true size of these quantities is actually 100 *times* larger again!**

A single $100 bill, our basic unit

One hundred $100 bills = $10,000

100 bundles of $10,000 bills = $1,000,000 ($1 million)

A pallet of 1,000 bundles of $10,000 bills = $100 million

10 pallets with $100 million on each = $1 billion

Remember that to get a real appreciation of how much a billion really is, we would need to use $1 bills, so the pile would be 100 times larger!

In the following image, compare how much larger a trillion dollars' worth of $100 bills is compared to the billion quantity that is in the lower left corner under the sign.

A thousand pallets worth $1 billion = $1 Trillion. This is an enormous number that is so large it's hard to conceive! A trillion dollars' worth of $1 bills would require 100 warehouses as large as this image!!

Very few people could even judge how many beads would be required to constitute a thousand. They typically think a thousand items is if far less than it is. This is why we humans have so little ability to go near visualizing large numbers. Even most scientists and mathematicians cannot do it.

Our number system uses a base of 10, probably due to us having ten fingers. When the units base is full, we move onto the next column and start again with tens, and then we do the same with the hundreds column and so on. In the original **old English version** of counting, ten tens was a hundred - a hundred hundreds was a thousand - a thousand thousands was a million - a million millions was a trillion – and a trillion trillion was a quadrillion, and so on. When this system was first considered, it was not visualized that anyone would need to use such large numbers as billions and trillions in the real world. In science, and now in finance, the English system became far too cumbersome to deal with, so the Americans devised the present more practical system that is now commonly used throughout the world. The **American system** increases by a thousand times (i.e. by the 3 zeros) in each progression, as shown above.

To help appreciate such large numbers, we can also use time as a way to comprehend large numbers. For example, imagine a tap dripping at one drop per second. At this rate, the tap would fill a bucket in 6.5 days. For it to fill an Olympic swimming pool, it would take a *billion* seconds or 31 years. To fill 1,000 Olympic pools, it would take a *trillion* seconds or 31,000 years! (These times assume no evaporation.) The Big Bang that created our universe occurred just under 14 billion years ago. This is considered an enormously long time. But a trillion years is 68 times as long as the universe has existed! As we can see, a billion is a very large number, but a trillion is such an enormous number that it is almost impossible to comprehend it. To nearly all people, billions, and certainly trillions, are just numbers that have no relativity to their everyday lives.

The debt levels across the world have risen so astronomically high that the biggest economies

	ORDERS OF MAGNITUDE		
10	Ten	has 1 zero, or 10^1	
100	1 Hundred	has 2 zeros, or 10^2	= ten tens;
1,000	1 Thousand	has 3 zeros, or 10^3	= ten one-hundreds
1,000,000	1 Million	has 6 zeros, or 10^6	= a thousand thousands
1,000,000,000	1 Billion	has 9 zeros, or 10^9	= a thousand millions
1,000,000,000,000	1 Trillion	has 12 zeros, or 10^{12}	= a thousand billions
1,000,000,000,000,000	1 Quadrillion	has 15 zeros, or 10^{15}	= a thousand trillions

measure their debt in trillions of dollars. To make it worse, national debts are fudged to make them look much less than they really are. The Global Financial Disaster of 2008 was caused largely by derivatives, which involved outright fraud at the highest level. The major banks have been carrying their worthless derivatives as valuable assets in their accounting records since 2008. These large banks have not written their derivative losses off as they should. If they did, they would all be insolvent. The Bank of International Returns calculated that the amount of money in the form of derivatives globally in the big banks is estimated to be $4.5 Quadrillion! This is an unimaginably large figure. On top of this, there is government debt and private debt which is also astronomically high. The world economy will soon have to deal with this and have a reset of all currencies, and when it does, it will very likely cause a worldwide depression far worse than that of the Great Depression in the 1930s. This could cause in excess of a billion deaths. Although we talk about numbers in astronomy being astronomical, those in the financial world are beyond astronomical in their size!

A LIGHT YEAR

In earlier times, when using terrestrial measurements like kilometers and miles for the sizes and distances to astronomical objects, astronomers had to deal with numbers even larger than billions and trillions. Using kilometers or miles to measure distances to the outer planets required billions, and the edge of our solar system required measurements in the *trillions*. With earthly measurements, distances to the stars were measured in *hundreds of trillions, to thousands of trillions,* and those to galaxies are *millions to billions of trillions!* To overcome using such enormously large, cumbersome numbers, astronomers cleverly came up with a new unit of measurement. They called it a ***light year***. It is the distance that light travels in one year. It is not a measurement of time: it is a measurement of distance. **One light year is almost 10 trillion km (6 trillion mi).** In digits, that's approximately 10,000,000,000,000 km. So rather than talking about trillions of kilometers to nearby stars, astronomers could now talk in distances of tens or hundreds of light years to most stars. This was much more practical.

THE SPEED OF LIGHT

In the vacuum of space, there is almost no matter, so there is almost no resistance to light traveling at it maximum speed.

Nothing in the universe can travel faster than light. Its velocity is a phenomenal 300,000 km every second (186,000 m/sec). Why it is that speed, no one knows. At this speed, a light beam will travel to the Moon in just over one second! Try to imagine that. However, because distances in the universe are so enormous, even light speed is incredibly slow when it comes to crossing distances between the stars (interstellar distances), let alone distances between galaxies (intergalactic distances). Even at the speed of light, it would take decades to go to most of the closest stars. More distant stars in other spiral arms of our galaxy would take thousands of years. To reach other galaxies would take millions to billions of years. Unfortunately, faster-than-light-travel is not an option because Einstein proved it is impossible to travel anywhere near the speed of light, no matter how technically advanced a civilization might become. The laws of physics forbid it.

The speed at which light travels is dependent of the density of the medium through which it travels. As the density increases, the speed of light decreases, as shown in the adjacent table:

Medium	Speed km/sec
Vacuum	300,000
Air at sea level	298,000
Water	225,000
Glass	200,000
Diamond	125,000
A Bose-Einstein Condensate*	0.7 to 0

THE SPEED OF LIGHT IN DIFFERENT MEDIUMS

*A Bose-Einstein Condensate is a recently created, super-dense form of matter that only exists at a mere one ten-billionth of a degree Kelvin above absolute zero! At this temperature, atoms no longer vibrate, as they do at higher temperatures. In a condensate, millions or billions of atoms clump together in the same spot to form a type of super atom. They all have the same energy level, so they have a single quantum wave. Recently, a light beam was made to come to a complete stop in a Bose-Einstein condensate! Micro-gravity in the space station is allowing astronauts to research the potential of this new, and very strange form of matter.

The Distance Light Travels In 1 Second

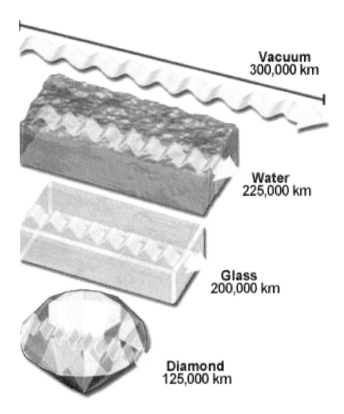

Part 2

PRACTICAL TIPS FOR STARGAZING

NAVIGATING YOUR WAY AROUND THE SKY

Just as we need a street map to find our way around a city, so do we need star maps to find our way around the night sky. There are many astronomy sites on the Internet that provide good star maps for use on a computer or tablet, but for the beginner standing under the stars, an astronomy smartphone App is the most practical. Apps are like having a pocket planetarium in your hand. They use global positioning satellites (GPS) to determine your location on Earth to show you exactly the sky you are seeing at your local time in the direction in which you are looking. You hold the phone up to the sky to match the screen image against the sky. On the Internet, there are videos about these Apps that show you their many features. Some are free while others are a mere $2.

A time slider allows you to see what's coming up later in the night, or any time in the future, or what the sky was like at any time in the past.

Apps for smartphones to identify objects in the night sky are especially practical and useful.

And you can also easily change the location to see what the sky looks like from other places around the globe.

By holding your smartphone up to the horizon in the direction you are looking, you can easily identify celestial objects in that part of the sky. As you change your direction, or the angle of your smartphone from the horizon to overhead, the sky on the App moves with your gaze. To see more detail in the sky, you simply widen your fingers to zoom in. These Apps have the ability to brighten or darken the sky on your screen to simulate the darkness of the sky at your site. You can simulate a naturally dark sky, one with city light pollution or moonlight, and even twilight.

Stars, constellations, the planets, the Moon, the Sun, and the Milky Way can be selected to appear or not appear, as well as thousands of deep sky objects. By tapping on an object, this will identify it, and it will bring up photos and descriptions of it. For some objects, you will also get its history. While these Apps are very practical for

Here are two constellations that are easy-to-recognize, Orion, the hunter (left), and the 'V' shaped Taurus the bull (right of center). The Seven Sisters star cluster is to the far right. Credit: Jerry Lodriguss

quickly identifying naked eye objects in the sky, the screen is often too bright to use if you are observing faint deep sky objects or comets. And if someone rings you while you are at your telescope using the App, the screen will go to full brightness and blind your night vision, so it's best to have the brightness down as low as is practical and to block incoming calls. These Apps are ingenious technology, given how practical they are.

The **Skyscout** is a compact computerized device that works like the Apps above. It has a tilt sensor and an electronic compass that computes what you are aiming it at. It can be used hand-held, on a telescope, or with a pair of binoculars. Via earphones, or by looking at the screen, you are provided with technical information about objects that you are pointing it at. It computes the location of the Moon and planets. Its backlit screen shows constellations. To identify an object, point the SkyScout at an object you are looking at, press the Target button, and it will identify it. When you look through the SkyScout's viewfinder, you will see illuminated rings. To find a specific object, select it from the SkyScout's menu and then follow the direction markers in the viewfinder. It costs around $700 brand new. It can be obtained secondhand for far less. YouTube has a good explanatory video on this product.

On the Internet, there are also very useful, feature-packed, free programs for your computer or iPad.

Celestron's SkyScout personal planetarium

Here we see the bright stars of the constellation, the Big Dipper, Ursa Major lying close to the horizon. Credit: Jerry Lodriguss

STAR ATLASES

If you are using binoculars, or a telescope that does not have computerized object-finding technology like Sky Scout, or a **GOTO system** (Go To this object), then a good star atlas is very useful for star-hopping to deep sky objects. **Norton's Star Atlas** is particularly good for finding your way around the sky and for naked eye observing of the constellations. It has the advantage of providing maps that cover large areas of the sky. This makes it easy for beginners to learn the constellations and to know where major deep sky objects are located.

There are many star maps online but it is hard to find one that is practical and accurate enough for serious observing at the telescope. **Wil Tirion's Sky Atlas** provides detailed star fields for telescopic observing. Due to its larger map sizes, there are more deep sky objects shown than in Norton's Star Atlas. Because Tirion's fields of view are smaller than Norton's, it is not as easy for beginners to use as Norton's is.

Both atlases are useful to have. They are money well spent. When using a star atlas, be sure to orientate it exactly as you see the star patterns in the sky. Trying to

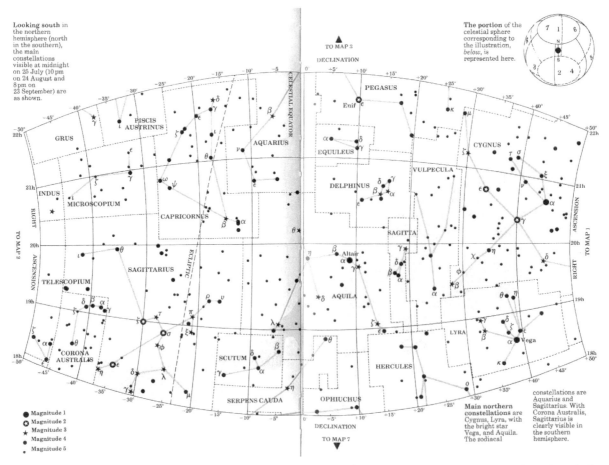

This is a map opening from Norton's Star Atlas. These maps show a large area of the sky. They contain all stars visible to the naked eye down to magnitude 6. The Milky Way is shown in light blue and constellations are marked with solid light blue lines. This atlas is good for using when observing with the naked eye in a relatively dark sky.

rotate them in your mind is not easy, so you can easily become confused. You can plot the positions of comets and asteroids in pencil on Norton's maps and then rub them out when you are finished with them. If you photocopy the charts, you can use these to sketch in the shape and length of the tails of bright naked eye comets as they change night by night.

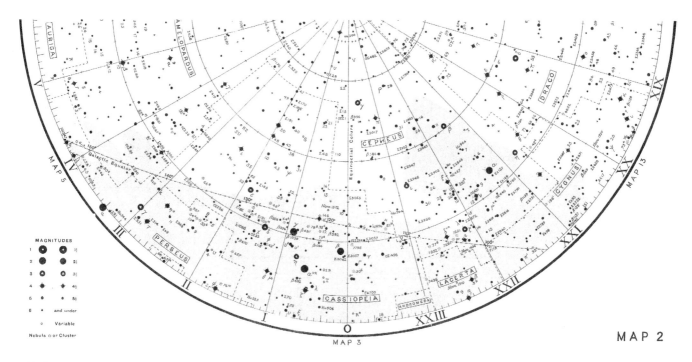

Half of Norton's circumpolar map for the northern sky in its older form.

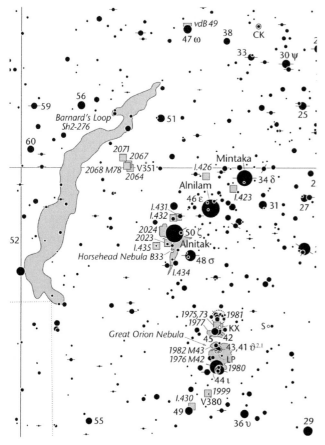

This is a section from Wil Tirion's Sky Atlas that shows a part of the constellation of Orion that includes Orion's Belt and his sword. These stars are often referred to as the saucepan. Stars to magnitude 8 are shown. These are visible in binoculars. This atlas is excellent for showing deep sky objects, most of which can be seen in amateur telescopes. Each type of object has its own symbol. This atlas depicts the shapes and relative sizes of the largest objects.

MEASURING THE SKY IN DEGREES

It can be useful to know how to measure or estimate the sizes of objects in the sky if you do not have a star atlas handy. If you have a perfect horizon all around, the sky from one horizon to the opposite horizon measures 180°. From a perfect horizon like the ocean, to directly overhead is half this i.e. 90°. The distance between Alpha and Beta Cassiopeia in the far northern sky, or the two stars that form the Pointers to the Southern Cross (Alpha and Beta Centaurus) in the southern sky, is approximately 5°. The distance between each of the three stars that form the belt of Orion is 1.5°. The Moon and the Sun both have a diameter of ½° or 30 arc minutes (30').

There are times when you need to measure the length of a comet's tail or a meteor trail. A simple way of roughly measuring lengths in the sky is to use your outstretched hand at arm's length. From the tip of your thumb to the tip of your little finger is about 20°. A clenched fist at arm's length is about 10° across. Most people's fingers usually cover about 1° of sky. Surprisingly, a fingertip is twice the diameter of the Moon or the Sun! Even your little fingernail is larger than the apparent diameter of the Moon or the Sun.

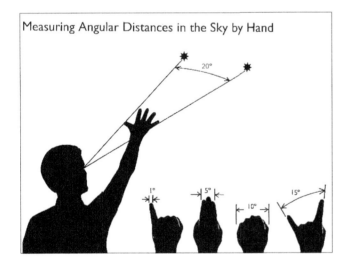

PART 3

COMMONLY USED ASTRONOMICAL TERMS

The following terms are frequently used in astronomy, so you should become acquainted with them. Astronomical names and words are shown in bold italics to make them easy to find.

MAGNITUDE – A SCALE FOR MEASURING BRIGHTNESS

Star brightness is expressed by the term 'magnitude'. Each step in magnitude is 2.5 times brighter or fainter than the adjoining one. The scale is such that a 6th magnitude star is 100 times fainter than a 1st magnitude star.

The faintest stars that a keen naked eye can just see in a dark, clear sky are magnitude 6. Under perfect conditions, a very experienced observer with extraordinary eyesight can see stars to almost 6.5, but only with averted vision and when sky glow is at its minimum.

The majority of bright stars in the sky are around magnitude 1 or 0. Stars that are exceptionally bright like Sirius, Canopus, Alpha Centauri, or the planets Venus, Mars, and Jupiter, particularly when they are at their closest to Earth, have negative (-) magnitudes. For example, Sirius is magnitude -1.5. Super-bright shooting stars can be -5 to -10. Very occasionally, some are many magnitudes brighter still. The Full Moon is -12, while the Sun is -27.

Stars that have magnitudes with a minus sign are therefore very bright while stars around magnitude 3 are of moderate brightness and those between 5 and 6 are very faint to the naked eye. In the centers of cities that are very brightly lit, you may only see stars to magnitude 3. In the inner suburbs of cities that have clean air, it is common to be only able to see stars to around magnitude 4.5. In a dark sky, you will see to magnitude 6 and by using common 7 x 50 binoculars you will show stars to at least magnitude 9.5. Under perfect conditions, a 100 mm (4") telescope will reveal stars to magnitude 12.5, and a 200 mm (8") telescope will see to magnitude 14. A good astrophotographer

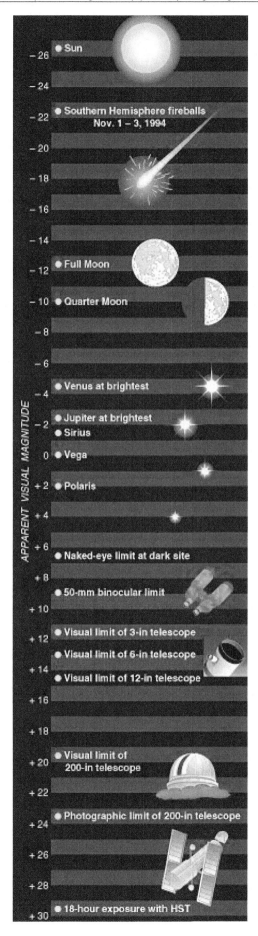

using a 310 mm (12.5") telescope can capture stars to magnitude 19, while the world's largest ground-based telescopes can photograph stars to around magnitude 29. Long exposures using the HST reach magnitude 30.

INTEGRATED MAGNITUDE

When determining an integrated magnitude for nebulous objects such as comets and deep sky objects, astronomers mathematically 'integrate' the object's total brightness, which is spread over an area, and treat it as an aggregate of all the light, thereby making it as if it is equivalent to a star-like point. Integrated magnitudes are a magnitude or two brighter than they appear to the eye because the object's light is spread out and therefore fainter when seen with the naked eye.

APPARENT MAGNITUDE

The *apparent* magnitude of an astronomical object refers to its brightness in comparison to other stars in the sky. When we look at stars with our eyes, we have no way of knowing whether a star appears bright because it is a small, faint star that is very close to us, or because it is a large, bright star that is far away. Because a star's apparent brightness is a relative term, it tells us nothing about the true, *intrinsic* brightness of the star.

The stars in this cluster, which are in the Small Magellanic Cloud galaxy, are roughly the same distance from us, so their intrinsic brightnesses are as we see them. Credit: NASA HST

INTRINSIC BRIGHTNESS

The term *intrinsic* brightness refers to a star's true brightness: this is quite different to its *apparent* brightness. The intrinsic brightness allows astronomers to compare stars as if they are all at the same distance. To determine the intrinsic brightness of a star, astronomers use special instruments on large telescopes called **spectrographs** to record their **spectrum** (rainbow colors). This tells them what chemistry the star has, and its dominant overall color. This tells us what its age is, its type, its temperature, its mass, its internal pressure, its surface gravity, and its diameter, as well as how fast it is moving towards us or away from us, and how fast it is rotating. Amazingly, all these things can be determined from the light of a star! Repeated observations measure how it may vary in brightness, which can tell astronomers what type of star it is. Estimates often have to be made because there are variables and unknowns that can affect a star's brightness, such as how much intervening dust lays between us and the star.

APPARENT DIAMETER

The apparent diameter of an object is the size it measures across the sky. This tells us nothing about the object's

During a total solar eclipse, it is easy to see that the Sun and the Moon have almost the same *apparent* diameter despite them having hugely different in *intrinsic* diameters.

real size. For example, the Sun's apparent diameter is the same size as the Moon's, but in reality, the Sun is almost 1.4 million km (0.9 million mi) in diameter, whereas the Moon is only 3,500 km (2,000 mi) in diameter, yet they appear the same diameter in the sky because they are not at the same distance from us. The Sun is

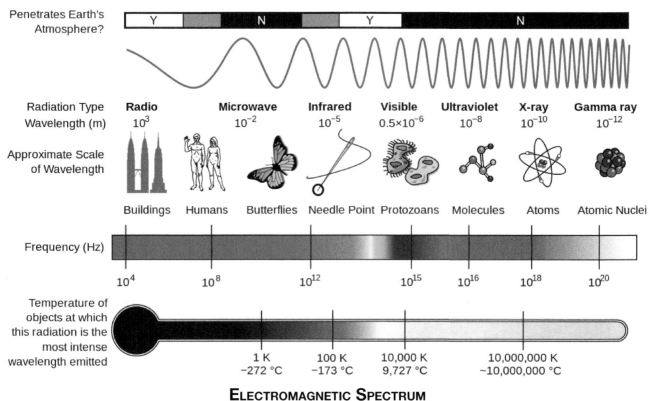

ELECTROMAGNETIC SPECTRUM

approximately 150 million km (100 million mi) away, whereas the Moon is only around 380 thousand km (240 thousand mi) distant. It is a coincidence that they appear to be almost the same apparent size.

THE ELECTROMAGNETIC SPECTRUM

The electromagnetic spectrum is the entire range of wavelengths or frequencies of energy in the universe.

It extends from the shortest and most energetic gamma rays to the longest radio waves. The visible light we see lies near the center of this spectrum. Earth's atmosphere does not allow the transmission of the frequencies that have a black band above them with the letter N, which equates to 'No transmission through the atmosphere'. Only space telescopes can observe these frequencies. Y = Yes, meaning that this frequency penetrates our atmosphere.

A BINARY STAR SYSTEM

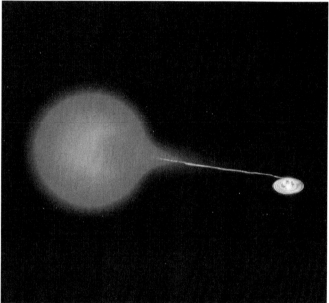

When two stars orbit one another, or more correctly, they orbit a common center of gravity, this is referred to as a binary star system.

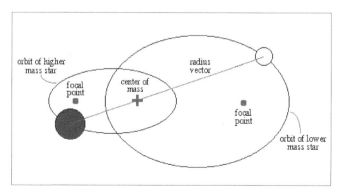

A binary star system has two stars gravitationally attached to one another, so they orbit a shared center of gravity, the location of which is based on their relative masses.

AN ASTRONOMICAL UNIT

An Astronomical Unit (AU) is the distance from the Sun to the Earth.

CONSTELLATIONS

A constellation is a particular group of stars that represents some form of imaginary figure or object: e.g. Leo the lion, Orion the hunter, or Libra the scales. There are 88 constellations. The original constellations were prominent patterns of relatively bright stars together with some fainter ones. They were named as far back as tens of thousands of years ago. Ancient cultures such as the Babylonians, the Chinese, the Indians, and the Greeks often had their own names for star patterns. The Greeks named 48 of the mythical figures used today for such constellations as Pegasus the winged horse, and Cygnus the swan.

People several millennia ago used the constellations for navigation at night over land and later over the sea. Constellations were used to refer to specific areas of the sky and they were incorporated into story telling by all cultures. All the ancient constellations that were recorded in writing, were those visible from the northern hemisphere. People in the far southern hemisphere had their own constellations, but they had not invented writing in those times to record them, so we do not know much about what they were. Constellations in the far southern sky were eventually named by the major cultures of Europe after European maritime explorers were able to sail into the far southern oceans to see them.

The far southern constellations were given their names by European astronomers in the 17th and 18th centuries. Astronomers of that time gave some far southern star groups Latin names for what in those days were exciting new technologies such as Antlia the air pump, Microscopium the microscope, and Circinus the compass.

In the 19th century, astronomers determined specific areas of the sky for the modern-day constellations. The borders of constellations were based on celestial longitude and latitude.

Here we see the constellations Orion to the left, the V shape of Taurus at center and the Pleiades near an overexposed New Moon at right. Credit: Jerry Lodrigus

The Pointers are the two bright stars to the left. They are Alpha and Beta Centaurus. They 'point' to the Southern Cross, which is the 4 bright stars in the middle known as the constellation Crux. The large pink Eta Carina Nebula (far right) which is visible to the naked eye, lies in the constellation of Carina. Credit: Akira Fujii

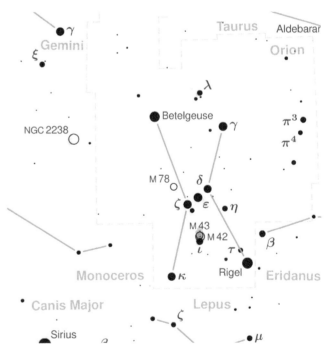

All constellations, such as Orion shown here, have designated boundaries as indicated by the yellow dotted lines that are determined by celestial longitude and latitude.

This view shows the constellations from Crux, the Southern Cross, through Centaurus, the centaur, to Libra the scales on the right. Venus, Mars and a New Moon are visible in the low west in this view taken for a planetarium App.

THE CELESTIAL POLES

The **North and South Celestial Poles are points in the sky where the North and South poles of the Earth point into space.** Due to the daily rotation of the Earth, the sky rotates around these points. The celestial poles are not coincident with the magnetic poles, which are around 10 degrees off center to the poles of the Earth's rotation.

LOCATING THE NORTH CELESTIAL POLE (NCP)

The 2nd magnitude star **Polaris** is commonly known as the **North Star**, or the **Pole Star** because it is situated almost on the NCP. In the south, there is no bright star to help locate the SCP; only a faint 5th magnitude star called **Sigma Octantis** lies close to it. When standing at the equator, the NCP lies exactly due north on the northern horizon while the SCP lies exactly due south on the southern horizon.

The Celestial Pole in either hemisphere is the same number of degrees above the horizon as the observer's latitude, i.e. it will be 30° above the horizon if your latitude is 30°. Stars in the northern sky rotate around the NCP from northeast to northwest while those in the southern sky rotate around the SCP from southeast to southwest.

Left: **Star trails rotating around the NCP are seen here over Mauna Kea. Polaris is the bright star almost at the center of the star trails.**
Credit: Peter Michaud (Gemini Observatory), AURA, NSF.

How to find Polaris, the star for the NCP. It is 5 times the distance between the two bright stars of Ursa Major that form the outside of the Big Dipper. The northern sky rotates around the pole star. Credit: Jerry Lodriguss

LOCATING THE SOUTH CELESTIAL POLE (SCP)

The Southern Cross and the Pointers can be used to determine the approximate location of the SCP. Due south is below this point.

To find the SCP, look south and locate the Southern Cross (if it is above the horizon at the time). If you imagine an imaginary line drawn through the long axis of the Southern Cross and extended 4.5 times that length, this will be approximately at the SCP. Another rough approximation is to draw an imaginary line at a tangent to the Pointers. Where this line crosses the line that goes through the long axis of the Cross is also close to the SCP.

Here we see star trails around the SCP. The colors have been exaggerated to bring out differences in the colors of the stars. Note that no bright naked eye star exists at the SCP. Credit: Lincoln Harrison

RIGHT ASCENSION & DECLINATION

Just as maps of Earth have longitude and latitude, so do maps of the night sky. In the sky, the equivalent of longitude is called Right Ascension (RA). The equivalent of latitude is Declination (Dec).

Just as the sky has a north and south pole, it also has a **Celestial Equator**. This is situated directly above the Earth's equator - half way between the celestial poles. Again, as with the Earth, there are 90° between the Celestial Equator and the Celestial Poles. The Celestial Equator is 0° Declination while the NCP is + 90° and the SCP is - 90°.

If we draw a concentric line around either pole at any distance from it, this describes a line of **Declination (Dec)**. Lines of Declination north of the Celestial Equator are given positive (+) numbers up to 90 while those south, are given negative (-) numbers. Each degree of declination is divided into 60 minutes of arc ('). The minutes are divided further into 60 seconds of arc ("). An accuracy of seconds of arc is generally only required for high precision telescopic measurements.

Just as lines of longitude on maps of the Earth are drawn from pole to pole, so are lines of **Right Ascension (RA)**. The sky is broken up into 24 'hour' zones of RA. These relate directly to the Earth's rotation each hour. The 24 hours of RA are divided into 60 minutes, and in turn, each minute is divided into 60 seconds.

If we know the RA and Dec coordinates of a comet for any night, then we can find its exact position in the sky. Let's take an example; say a comet is at RA 14hr 27m, Dec -41° 56m. Using a star atlas, simply find the hour of RA (14) and then the minutes (27). Run along this line until it crosses the Dec line at -41° and 56 minutes. Where the lines cross, is the position of the comet's head. As a test, use a star atlas to see what famous object lies at the following co-ordinates: RA 00 hr 40", Dec +41° 00".

Modern computer-aided telescopes have a GOTO (Go To) facility whereby motors on both the RA and Dec axes drive the telescope to the coordinates of the object you select.

DIURNAL ROTATION

Diurnal rotation is a term that refers to the apparent movement of the stars across the sky caused by the Earth's rotation. Stars in the sky rotate around the celestial poles in increasingly larger circles the further they are away from the pole.

When looking east at stars lying along the celestial equator (like the three bright belt stars in Orion) these stars rise along fairly straight paths, whereas those near the poles travel along circular paths. As an example, the rotation of the Earth takes the Big Dipper in Ursa Major, and the Big 'W' of Cassiopeia in the northern sky, in a large circle around the NCP. In the southern hemisphere, the Southern Cross, the Pointers, and the Magellanic Clouds circle the SCP. At latitude +35° in the northern hemisphere, the Big Dipper stars only set below the northern horizon for a relatively short time before they rise again. Similarly, at latitude of -35° in the southern hemisphere, the Southern Cross only sets for a short time before it rises again.

The following stunning, super wide-angle, whole-sky image on page 78 shows spectacular star trails scribing arcs that become larger the greater their distance is from the celestial poles. When we look towards the east, the circles have become so large that the stars rise in straight lines from the eastern horizon passing overhead and setting 12 hours later in the west.

This amazing image shows the diurnal motion of stars as they move from the east to the west in arcs around the North Celestial Pole.

The photograph was taken using a very wide-angle lens and a time exposure over a whole night. It shows how the stars appear to move around the NCP when, in fact, it is the Earth that is rotating. This picture was taken at Mauna Kea Observatory from near the Gemini North Telescope, seen here in the foreground. The red beams are the trails of lasers that point to an object and then follow it while it is under observation. The laser beams follow the object to measure air turbulence along the line of sight to the object. Micro-adjustments can be made to the telescope's main mirror to counteract poor seeing. New beams occur each time the telescope moves to observe and track a new object. Credit: ESO

This wide-field image of star trails shows stars curving around the north celestial pole outside the top right and around the South Celestial Pole Beyond the bottom left. The stars passing across the sky in an almost straight line are on the celestial equator. The bright region diagonally across the center is the Milky Way. The vertical dashed line is the trail of an aircraft's wing strobe lights. Credit: Trenchard Photography

This is a 6-hour star trail image of stars rotating around the NCP above the Teide Observatory on Tenerife Is in the Canary Islands. Note the different colors of the stars and that Polaris is the star almost at the center of the circles. Credit: Baetosz Wojcznski

This image records stars trailing around the South Celestial Pole for several hours above a salt lake in the Atacama Desert in Chile. Credit: M. Duro

Star trail photography starting at the end of twilight and with an interesting foreground, can look very impressive.

CULMINATION

When an object reaches its highest point in the sky, it is said to have culminated. At culmination, an object passes across a line that joins the north and south celestial poles.

THE ZENITH

The Zenith is the term given to the region of sky directly overhead. It has a declination equal to the observer's latitude. For example, the Zenith Declination for London is +51° while Sydney or Buenos Aires is -35°. Objects with a declination equal to one's latitude will pass directly overhead when the object is at its highest point from the horizon - its **culminating point.**

THE ECLIPTIC

The Ecliptic is the path that the Sun takes as it travels around the celestial sphere throughout a year. The Ecliptic intersects the celestial equator at an angle of 23.5 degrees in a plane that passes through the centers of the Earth and the Sun. The **Celestial Equator** lies directly above the Earth's equator.

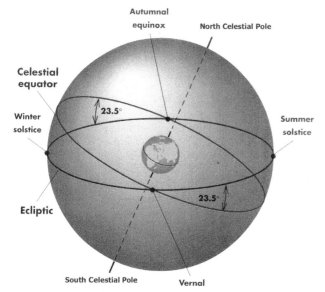

Astronomy: Journey to the Cosmic Frontier. Copyright 1995, Mosby-Year Book, Inc

THE ZODIAC

The Zodiac is a band of sky around 9° wide on each side of the Ecliptic. It defines the northerly and southerly limits of the paths that the planets, the Moon, and the Sun take as they move around the celestial sphere. The term 'Zodiac' literally means 'circle of the animals' because, in antiquity, many of the Zodiac constellations were represented as animals. The Zodiac is divided into 12 mythological star signs.

Some modern-day zodiac constellations extend beyond the edges of the Zodiac. Due to the wobbling of the Earth's axis around a circle known as its obliquity, this causes the equinoxes to move around the zodiac over 41,000 years. (See Volume 2 Chapter 1, page 2 for details on obliquity.) Due to this, astrological star signs no longer coincide with present astrological or astronomical star charts. This is one of many reasons why astrological forecasts are meaningless.

THE EQUINOXES

At the time of the equinoxes, we have equal day and equal night i.e. 12 hours of both. The equinoxes occur on March 20 and September 22. This is when the Sun crosses the Earth's equator and moves into the opposite hemisphere.

SOLSTICE

There are two solstices each year. The **Summer Solstice** occurs in mid-summer when the Sun reaches its highest point in the sky. It is the longest day of the year with the shortest night. The **Winter Solstice** occurs in mid-winter when the Sun is at its lowest point in the sky. This is the shortest day of the year with the longest night.

CONJUNCTIONS

Conjunctions occur when two or more planetary bodies, including the Moon, come close together. The term is most often used when referring to two or more naked eye planets coming close together, or when one or more of the planets are close to the Moon. Conjunctions of multiple bodies typically occur in the evening or pre-dawn twilight sky. This is because Venus and/or Mercury are often involved. Since their orbits are inside the Earth, this keeps them close to the Sun and therefore low in the western sky after evening twilight, or in the eastern sky before morning twilight.

ELONGATION AND CONJUNCTION

When a superior planet is on the same side of the Sun and at its closest point to the Earth, it is referred to as being at **opposition** with the Earth. At this time, it is at its highest point in the sky at midnight, and because it is at its closest to Earth, it is at its largest apparent size. The superior planets are said to be in **conjunction** with the Sun when they lie on the opposite side of it relative to Earth. This places them at their furthest point from us. They are not visible at this time due to the Sun's glare. These terms also refer to asteroids and comets.

When the inferior planets are at **inferior conjunction**, they are at their closest to Earth. They are not visible at this time because they lie beside the Sun in its glare. At this time, they present their dark side to us. Sometimes they pass directly in front of the Sun at inferior conjunction. When they lie on the far side of the Sun, they are said to be at **superior conjunction**. At this time, they are behind the Sun so they are not visible.

The term **greatest elongation** refers to the times when Mercury or Venus are at their furthest point away from the Sun, as seen from Earth. For Venus, it is between 45° and 47°. For Mercury, it is between 18° and 28°. When

A morning twilight conjunction of the planets, Venus, Jupiter, and Mercury (from top to bottom) as seen over one of the Very Large Telescopes at Paranal Observatory, Chile. Such conjunctions only last a few days at most before the planets drift apart as they move along their individual orbits. Credit: ESO

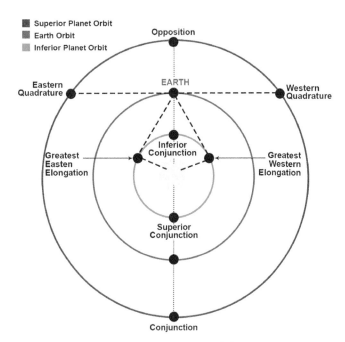

The *superior planets,* which are Mars and the gas giants Jupiter, Saturn, Uranus and Neptune, orbit outside the Earth's orbit, whereas the inferior planets, Mercury and Venus orbit inside it.

this occurs in the evening sky, it is referred to as them being at their greatest western elongation from the Sun. In the morning sky, they are at their greatest eastern elongation. (Refer to Volume 2 Chapter 7, page 209.)

PERIHELION - APHELION

Perihelion is where an orbit is the closest point to the Sun. **Aphelion** refers to the point at which the orbit is at its furthest from the Sun.

DEEP SKY OBJECTS

Deep sky objects are gaseous diffuse nebulas, planetary nebulas, open star clusters, globular star clusters, and galaxies. (See Volume 3 Chapter 4 through to Chapter 8.)

The brightest, and typically the largest deep sky objects can be seen with the naked eye as mostly faint, blurry, star-like glows. When observed in a telescope, these objects reveal surprising detail.

Common Names for Deep Sky Objects

Both professional and amateur astronomers have given **common names** to outstanding deep sky objects. Examples are the 'Whirlpool Galaxy' and the 'Tarantula Nebula'. These names describe an object's general appearance. Common names are preferred to having to remember catalog numbers. Because there are hundreds to thousands of objects in every object class, common names are only used for the largest and brightest objects, or those with distinctive features.

These photographs are examples of the different types of deep sky objects. From left to right are a diffuse gaseous nebula, a planetary nebula, an open star cluster, a globular star cluster, and a galaxy.

MESSIER (M) OBJECTS

Once you start observing the most impressive deep sky objects, you will come across **Messier objects**. They have designations like M4 – being the 4th object listed in **Messier's catalog of non-stellar objects**. Messier's catalog lists the brightest deep sky objects visible from the northern hemisphere. They have become popular classics for amateur astronomers to observe.

Charles Messier, a comet hunter in Paris in the late 18th and early 19th centuries, cataloged all the bright nebulas, star clusters, and galaxies that he and his comet-hunting contemporaries could see visually in their small telescopes that had apertures ranging from 90 mm (3.5") to 190 mm (7.5"). All but some of the brightest star clusters appeared as nebulous objects. These are

Messier 1 is a supernova remnant. Credit: NASA HST

now known collectively as **deep sky objects**. Messier's catalog helped him remember them so he didn't keep confusing them with comets that he was searching for. His catalog listed 103 objects. There were another seven deep sky objects in the northern sky that should have been bright enough for him to include, but curiously, he and his fellow observers did not list them. Several decades ago, these were included to make a total of 110 Messier objects.

In Messier's first catalog published in 1774, only 17 of the 45 objects in his catalog were discovered by him. In his final catalog, he discovered 40 of the 103 objects. Living in Paris, Messier could not see very far into the southern sky, so he was unable to include the largest and brightest southern deep sky objects.

🄲 During Messier Marathons, which are all-night star parties, stargazers try to observe all Messier objects in one night! Most Messier objects are impressive sights in today's large aperture, high resolution amateur telescopes that far surpass the telescopes in Messier's day.

Messier would have been impressed to know that his list of 'nuisance objects' would have made him famous in the late 20th century, especially as they were of little interest to him and other astronomers of his time, mostly because they did not know what they were.

NGC & IC OBJECTS

Deep sky objects with an NGC designation are those from the **New General Catalog.** It lists nearly 8,000. This catalog was originally compiled by **John Dreyer** in 1888. Most of these objects were from the copious observations made by the great astronomer **William Herschel** (the discoverer of Uranus) and his talented astronomer son, **John Herschel**. Between the 1960's and 1970's, the catalog was revised and improved upon by **Selentic and Tiff**. They removed errors in the NGC catalog and published it as the **Revised New General Catalog** (RNGC). This was later expanded into two Index Catalogs (IC) which included another 5,386 objects. **IC objects** were nearly all faint, distant galaxies. Numerous catalogs now exist for specific types of objects in every class, but these are generally only of interest to professionals and very serious amateur astronomers who specialize in observing certain types of typically very faint objects.

THE DARK OF THE MOON

The 'Dark of the Moon' refers to a period of time when the Moon is not visible in the sky. This occurs

Maximum detail is seen along the Moon's terminator. Credit: Steve Hill

before the Moon has risen at night, or after it has set. The longest period of the dark of the Moon is around New Moon when the Moon is close to the Sun. At this time, the Sun and the Moon both rise and set at similar times. The dark of the moon is the best time to observe deep sky objects.

TERMINATOR

The term **'terminator'** refers to the region of a planet, a moon, an asteroid, or a comet's nucleus where the day side meets the night side. This region shows the most surface detail because the Sun is very low on the horizon causing shadows to be cast from even very subtle changes in elevation.

ALBEDO

Albedo refers to the average reflectivity of a body. The Earth's white clouds and land masses reflect, on average, around 30% of the sunlight that falls on our planet, whereas the Moon's dark charcoal-gray surface reflects on average only 12% of the sunlight it receives. The albedo of a body can change dramatically if it has both very dark and very bright surfaces like Saturn's moon Iapetus has. When Iapetus is on one side of Saturn we see mostly its brilliant white hemisphere, so it is easy to see in amateur telescopes, but when it moves to the other side of Saturn, its very dark side faces us, so it becomes invisible. (See Volume 2 Chapter 10, page 384.)

DWARF PLANETS

'Dwarf planet' is a new term to describe planetary worlds in orbit around the Sun that are over 1,000 km in diameter and round in shape, but not massive enough to have enough gravity to have cleared the region of their orbit of other major asteroidal material. Pluto has been designated a dwarf planet because it fits these criteria. Large numbers of other dwarf planets are continually being discovered beyond the orbit of Pluto in the huge Kuiper Asteroid Belt. Ceres, the largest asteroid in the Inner Asteroid Belt between Mars and Jupiter has also been designated a dwarf planet because it fits the criteria. (See Volume 2 Chapter 12, page 438.)

METEORS & METEORITES

A meteor is a piece of debris in space that burns up high in the Earth's atmosphere. They are commonly referred to as **'shooting stars'** or **'falling stars'**. If they are very bright, they are called **bolides** or **fireballs**. Meteors are typically specks of dust, sand, or rocky pebbles left over from the formation of the solar system. When comets pass by Earth's orbit, they leave behind trails of dust-like debris. When Earth passes through this debris, it burns up in the atmosphere creating a **meteor** shower.

A meteorite is a meteor large enough to pass right through the atmosphere and hit the Earth's surface. (See Volume 2 Chapter 2.)

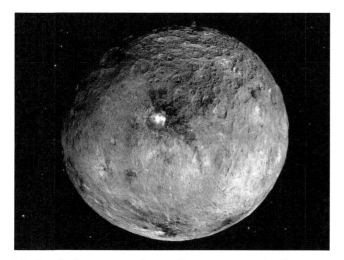

The dwarf planet Ceres orbits in the Inner Asteroid Belt. Credit: NASA

A Leonid meteor burning up in the atmosphere. Credit: Navicore

The dwarf planet Pluto orbits beyond Neptune in the Outer Kuiper Asteroid Belt. Credit: NASA

A piece of a large meteorite recovered in Western Australia. Credit: A. Oats

OCCULTATIONS

An **occultation** occurs when one astronomical object is wholly or partially hidden by another object that passes in front of it. When the Moon passes in front of the Sun during a solar eclipse, this is a special case of an occultation. Stars and planets often go behind the Moon's limb, either on its bright side or its dark side. This is referred to as a **Lunar Occultation**. The exact timing of these events can accurately determine the position of the Moon. When stars graze the edge of the dark side of the Moon, this is called a **Grazing Lunar Occultation**. A star can disappear and then reappear a few times as it goes behind lunar mountains and then reappear in the valleys between them. (For details see Volume 2 Chapter 5, page 131.)

Occultations by asteroids and dwarf planets occasionally occur when stars pass behind them. When several accurate timings are made at different sites on Earth of a star disappearing and reappearing behind an asteroid or dwarf planet, the dimensions and shape of the body

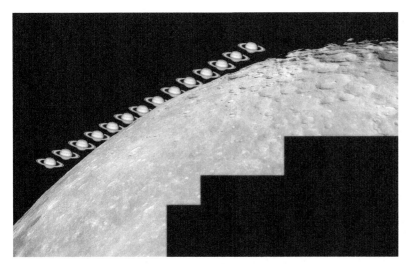

Left: **A grazing occultation of Saturn passing behind the limb of the Moon.** Credit: Peter Lawrence

Right: **A binocular view of an occultation of Venus by the Moon.** Credit: Amando Lee

A telescopic view of Venus prior to being occulted by the Moon Credit: Daniele Gasparri

can be determined fairly well. If the star dims before it disappears, and it does not brighten immediately after it reappears, then this is evidence of an atmosphere around the occulting body. Occultations of stars in globular clusters have been used unsuccessfully to search for stellar mass black holes.

TRANSITS

A transit is the passing of one object across the face of another object. Both Jupiter and Saturn's moons transit their respective planets. Occasionally, Mercury and Venus transit the face of the Sun.

The transit of a planet across the face of its star is used to discover other planetary systems. As a planet transits its star, it causes the light of its star to dim minutely for the period of its passage.

PHASES

The term 'phase' refers to the shape of the illuminated portion of the Moon or a planet. The **New Moon** phase is a crescent. The half-moon phase in the early evening is referred to as **First Quarter.** When it is a half Moon in the morning sky, this is referred to as **Last Quarter.** Because Mercury and Venus orbit inside the Earth's orbit, they display phases just like the Moon does. Mars has phases between full and three quarters. This is because it orbits outside the Earth's orbit, therefore the Sun illuminates it almost fully most of the time. Only when Mars is at an angle of about 90° relative to the Earth does it display a ¾ phase.

REFRACTION

Refraction is caused by the bending of light waves when they travel from one medium to another. When light passes from a vacuum into a medium with some density such as air, water, or glass, the light waves bend because they slow down. When light traverses the vacuum of space, it travels at its maximum speed, however, when it enters a dense medium, it slows down due to interactions with atoms in the medium. This compresses the light waves and causes them to change direction when they enter the medium at an angle. Refraction of sunlight in dense air at the horizon causes the Sun to rise about 2 minutes earlier and to set around 2 minutes later than it would if there was no air.

When light is refracted upon entering a denser medium, this can be likened to marching soldiers turning a corner. The closer a soldier is to the inside of the turn, the smaller his steps are and the slower he walks. The further a soldier is from the center, the longer his steps have to be to stay in a straight line and the faster he has to walk.

Top: **Here we see a transit of a Jovian moon and its shadow across the face of Jupiter. Such events occur regularly.**

Middle: **A transit of Saturn's largest moon and its shadow across Saturn's cloud tops. This only occurs every 7 years when the rings are close to edge-on.**

Bottom: **This is an illustration of a planet transiting the face of a dwarf star.**

Phases of the Moon throughout the lunar month

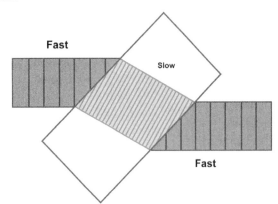

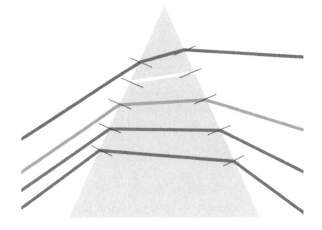

The refraction of light

Top: Light rays slow down when they pass into a dense material (the rectangle) and conversely, they speed up when they go back into a less dense medium. This change of speed causes light rays to change direction and be refracted.
Bottom: *Note that the more purple or blue the light is, the more it is refracted compared to longer wavelengths in the red end of the spectrum. When sunlight passes through raindrops, they act something like a double prism. The Sun's white light is refracting and split into prismatic rainbow colors when it passes into the drop. It then reflects off the back of the drop, and then passes through the front side of the drop on its way out. Again, the light is dispersed more creating primary colors of a rainbow.*

TWILIGHT

Twilight refers to the time when the sky is between being fully bright and fully dark. Evening twilight occurs after sunset and morning twilight occurs before dawn. Twilight is the time when there is some light from the setting or rising Sun reaching the upper atmosphere and coloring it. At twilight, the sky can be pink and purple while clouds can be colored yellow, orange, pink, and red. It is a peaceful time when birds sing and a sense of change invades the air. Many people think it is the best time of the day.

There are 3 forms of twilight: **1. Civil,** which ends when the Sun is 6° below the horizon, **2. Nautical,** which ends when the Sun is 12° below the horizon, and **3. Astronomical** when the Sun is 18° below the horizon. This is when the sky is fully dark. At the equator, the Sun sets perpendicular to the horizon, so it sinks very fast causing nautical twilight to last only 20 minutes. It is jokingly said that night 'falls' so fast at the equator that it is like the dropping of a blind. At high Arctic latitudes, the Sun sets at a very shallow angle to the horizon. In summer, even at midnight, it remains just below the horizon, so twilight lasts all night. In mid-winter, for 2-3 weeks, it does not rise so even at midday it is fairly dark, so civil twilight lasts all day. Twilight is at its shortest at the equinoxes i.e. when there is equal day and night.

Twilight is the time when the Sun is just below the horizon and the upper atmosphere is still lit by the light of the Sun.

SCINTILLATION

The twinkling of stars is technically known as 'scintillation'. It is caused by turbulence in the atmosphere. Some layers of the atmosphere move in different directions and others have different temperatures to one another. These two factors cause turbulence, and this refracts starlight at different angles making stars appear to change in brightness and in color. When a star or planet is seen in a telescope, turbulence in the air makes them appear to jiggle around and look blurry.

Scintillation is particularly noticeable to the naked eye when stars are close to the horizon. There, starlight passes through varying densities of air between the upper and lower layers. There are also differences in temperature. These two factors determine how much a star's light is bent or refracted. Low in the sky, there are also differences in transparency due to variations in water vapor and impurities such as dust, smoke, pollen in different layers of air. These factors can make a star's light change randomly in brightness. Refraction of starlight also makes stars change color. This is due to the atmosphere acting like a prism to disperse a star's light into its main primary colors. We see the blue end of the star's spectrum for a fraction of a second and then the red end. When there is a lot of scintillation, this is called **poor seeing**. (See Chapter 6, page 96.)

NORTH, SOUTH, EAST & WEST

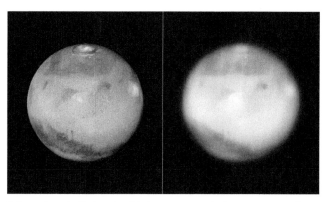

When there is little scintillation, stars and planets appear sharp in a telescope (left) but when scintillation is strong, this produces poor seeing so detail disappears (right).

It is important to know where the **cardinal points** are when we tell someone to look towards a particular direction. To determine **north, south, east, and west,** you can use a traditional compass, or a digital one on your watch, or the one on your smartphone. To find east without technology, east is in the direction that the Sun rises at the equinox, and west is where the Sun sets during the equinoxes, which occur on March 21, or on September 23. To find south, look directly east then outstretch your right arm at 90°. Where it points is due south. North is where your left arm points at 90°.

AIRGLOW

Airglow, or skyglow, is faint light that emanates from the top of the atmosphere. It stops the night sky being pitch black. It is caused by particles in the solar wind from the Sun and cosmic rays from deep space colliding at high velocities with air molecules in our upper atmosphere. This causes them to fluoresce faintly. In a naturally dark sky, when your hand or trees are silhouetted against the night sky, they appear black and the skyglow appears gray. If there was no air, the sky would be black like it is on our airless Moon.

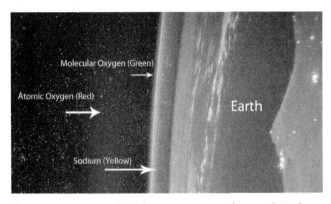

There are two types of airglow, one green and one red. Both are too faint for us to see their color. The greenish airglow occurs in the very uppermost thermosphere at around 150 km above the Earth's surface. It can look brownish due to sodium atoms in the air below it giving off faint yellow-orange light. The reddish airglow occurs at an altitude of 250 km. Nitrogen and oxygen air molecules fluoresce due to solar wind particles hitting them. Airglow is not evenly distributed due to the solar wind varying in its intensity. Airglow often appears rippled in photographs.

In photographs, both green and red airglow can be seen to form waves across the upper atmosphere. Airglow varies in brightness during the night and due to how active the Sun is. Credit: Dave Lane

Chapter 6
LET'S GET PRACTICAL

Supernova searcher and astrophotographer Greg Bock on the Gold Coast in Australia, observes from his large deck. This photograph was taken under the light of a Full Moon. Credit: Greg Bock

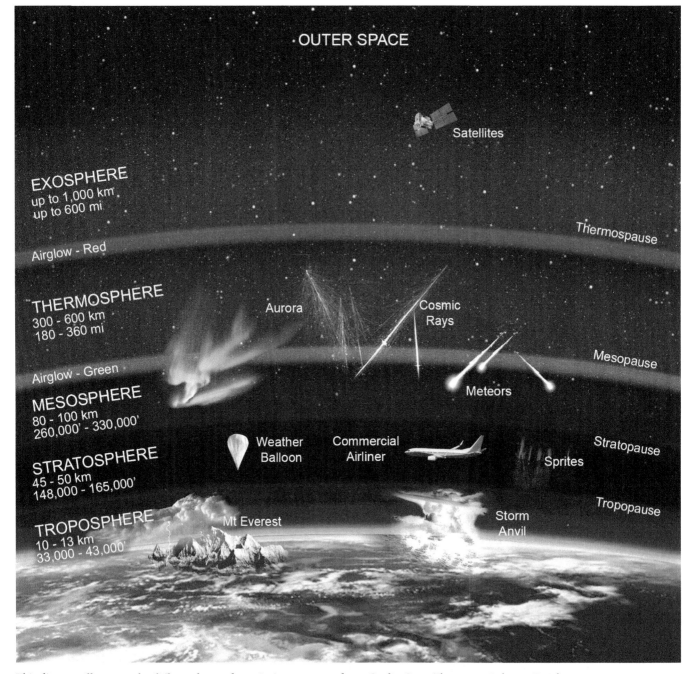

This diagram illustrates the different layers that exist in our atmosphere. Credit: Gregg Thompson, Rebecca Gjerek

In this chapter, we will focus on gaining a basic understanding of the structure of our atmosphere and how the weather works, because this plays a big role in observational astronomy. We will then look at things that make stargazing more practical, comfortable, and enjoyable. Comfort and convenience are essential because the more comfortable and relaxed you are, the longer you will want to observe.

UNDERSTANDING THE ATMOSPHERE

To understand how the weather affects observing conditions, it's important to have a basic appreciation of how the atmosphere works. Some of the answers to common questions about observing in this section may surprise you.

Our surprisingly thin atmosphere and the Earth's magnetic field are what protects us from being bombarded by rocks and radiation from space.

We will now look at how the atmosphere is structured and how it affects how clearly we see objects in space.

THE STRUCTURE OF THE ATMOSPHERE

The **troposphere** is the lowest layer of the atmosphere sitting above the surface. It is where all weather occurs, so it is very dynamic. The **stratosphere** is where commercial aircraft fly. It has a stable airflow. Above this is the **mesosphere** where auroras and meteors occur. It too is very stable. Beyond this is the extremely rarefied **thermosphere** where red and purple high altitude auroras occur. It is also where cosmic rays start to form showers of particles. The outermost layer is the extremely rarefied **exosphere** in which the Hubble Space Telescope and the International Space Station orbit together with thousands of satellites, both functional ones, and old disused one orbit. This layer fades away to nothing at about 1,000 km. There is a region of the upper atmosphere that contains charged atoms. It is referred to as the **ionosphere**. It extends from 85 km to 700 km (50 - 400 mi). It reflects long wave radio signals.

THE TROPOSPHERE

The **troposphere** extends up to 6 km (20,000') over the North and South Poles and up to around 10 to 13 km (30,000' - 43,000') over the equator. The Sun heats the air most over the equator making it expand to form the 'troposphere bulge' there. As the Earth's rotation is at its fastest at the equator, this also makes the air bulge out there. At the equator, the Earth is rotating at 1,600 km/hr (1,000 mph) whereas at the poles it is zero. At any location on Earth, the atmosphere's height will fluctuate due to changes in temperature. This is what causes atmospheric pressure to change.

The troposphere contains more than 85% of the air and 99% of the atmosphere's water vapor, so this is where nearly all the weather takes place. Plant life, and nearly all animal life, only survives in the lower troposphere up to an altitude of 3 km (2 mi) above sea level. Above this height, it is very difficult for life to survive because it becomes very cold, very dry, and there is insufficient oxygen.

Temperature in the troposphere decreases at a rate of about 2°C with every 300 m (1,000') of altitude, but at the top of the troposphere, the air stops cooling.

THE STRATOSPHERE

The layer above the troposphere is known as the **stratosphere**. It extends up to around 50 km (~ 30 mi or 160,000') above the troposphere. Where the troposphere and the stratosphere meet, this is called the **tropo*pause***. There is only minor turbulence here. This is where high altitude cirrus clouds form. Commercial airliners fly in the lower stratosphere. It warms with increasing altitude and it becomes increasingly dry as there is hardly any water vapor. What little there is, is in the form of tiny ice crystals that form cirrus clouds. At mid-latitudes, winds in the stratosphere typically move at around 200 km/hr from the west to the east.

THE MESOSPHERE

Above the stratosphere lies the relatively thin **mesosphere**, which extends from 80 km to 100 km (260,000' - 300,000') above sea level. The temperature here decreases with height dropping to -90°C (-130° F) making it the coldest layer. Green-curtain auroras occur in the mesosphere and meteors start to burn up here. Small quantities of hydrogen and helium exist in this layer.

THE THERMOSPHERE

Above the mesosphere is a layer called the **thermosphere**. It's called this because air molecules and atoms here are the 'hottest' reaching 1,200°C. But the air here is so thin that it is close to a vacuum. This layer extends to over 600 km (360 mi) high! Red and purple auroras occur here.

THE IONOSPHERE

In the upper thermosphere lies the top of the **ionosphere**, which contains charged atoms. This layer is used for bouncing long wave radio waves around the globe. Beyond this layer lies the **Van Allen radiation belts** that protect Earth's surface from much of the harmful radiation from the Sun, and also from a substantial percentage of cosmic rays from deep space.

The Exosphere

The exosphere is really the upper thermosphere, which consists of extremely rarefied atoms of air that fade away into space. Satellites orbit Earth in the **exosphere**. It is blown away downstream of the Earth by the Sun's solar wind. It is thought to extend for possibly as far as 100,000 km (60,000 mi).

HOW THICK IS THE ATMOSPHERE?

We are largely unaware that the biosphere, where all life exists, is only a thin film of air just 3 km in depth. Its thickness is 4,300 times smaller than the diameter of the planet! That's equivalent to a balloon half a meter across (20") being covered in a single film of cling-wrap! Most life lives at the bottom of this thin film.

Because our atmosphere is so thin, it provides only minimal resistance to prevent the Sun's heat from being lost to outer space. Fortunately, the oceans act as a heat sink to stop the dark side of Earth freezing over every night when it receives no warmth from the Sun. Water vapor in the air helps to retain some of the daytime heat. When low clouds are around, they act as a temporary 'blanket' to help retain some of the surface heat. Exactly why this happens is not understood. In winter, when water freezes out of the air making it dry, surface heat can quickly radiate away into space making it very cold if there is no cloud layer. Temperatures in deserts can be 50°C during the day but because there is no moisture in the air, at night the temperature can drop to below freezing shortly after the sun sets. This demonstrates that the minute trace of CO_2 that is in the air does not retain any heat.

At temperate latitudes, there is usually plenty of water vapor, so temperatures typically drop at night only by an average

In this picture taken from space, the troposphere and the stratosphere appear as the thin blue haze that is barely visible on the horizon. All life on Earth lives within 3,000 m of the lower troposphere. Our atmosphere protects life from the Sun's ultraviolet rays, most cosmic rays, and the cold vacuum of space. Credit: NASA

To appreciate how thin the atmosphere is, imagine the Earth shrunk to a sphere 50 stories high. At this scale, the troposphere would be only 40 mm (1.5") thick! It would be invisible in this view. At this scale, humans live within only 10 mm (½") of the surface. We are protected from the dangers in space by a very thin layer of nitrogen and oxygen gas, and the Earth's magnetic field. Credit: Gregg Thompson

of 10°C. In very cold regions, it can drop 20°C at night. When sunbathing, it is easy to notice the warming effect of sunlight because there is an immediate drop in temperature when the Sun goes behind even a small, thin cloud.

The Sun is responsible for 99.99% of all temperature on the Earth's surface so any change in solar activity produces a change in global temperature on the Earth, as well as on all the other planets and their moons.

What Is the Composition of the Air?

Nitrogen makes up a little over 78.09% of the air and oxygen is 20.95%. The inert gas argon constitutes 0.93%. Carbon dioxide (CO_2) is a trace gas being only 0.04% of the air. Only 3% of the 0.04% is made by man. That's a trivial 0.00005% of the air. By never telling the people it is so little, the media hysteria about this gas has led people to believe that there is hundreds of thousands to a million times as much in the air. The balance of trace gases which make up 0.07% of the atmosphere are methane, helium and hydrogen. Nitrogen, oxygen, and carbon dioxide are essential for life.

CLOUD TYPES

An internally lit storm cell from lightning with a perfect anvil.

High altitude cirrus clouds, commonly called horsetails, move across the sky from west to east in a jet stream at around 200 k/hr. The often foretell of a change that brings rain. When cirrus clouds are around, the seeing is typically poor. Credit: Gregg Thompson

Left: **This image captured a nearby, rapidly-growing storm cloud. Video of this shows the edges billowing out and being drawn back into it as hot air rises inside the cloud making it expand and rise higher. Sometimes developing storm cells build into a large storm and other times they dissipate once their top passes into the stratosphere.** Credit: Gregg Thompson

Stargazers benefit by having a basic knowledge of meteorology because this allows them to understand what the weather is doing so they know whether they will have clear skies or good seeing. There are many websites that explain the basics of meteorology. Here, we will just deal with those aspects that affect observing.

Clouds can interfere with observing so it helps to know the main types.

Low fluffy clouds that form close to the surface of the Earth extend up to 2,000 m (6,500'). They consist of fine water droplets (mist) that collect together to produce rain showers when they pass into cold air. Middle-level clouds form above 2,000 m and extend up to 6,000 m (20,000'). Typically, these do not produce much rain, unless it is a thunderstorm. High altitude cirrus clouds above this height consist of fine ice crystals that typically form in a stream of steady air in the lower stratosphere.

Cirrus clouds often, but not always, foretell of a change that involves rain coming in a day or two. In the diagram on page 96, the storm clouds on the right extend through all three levels. They typically produce rain, hail, lightning, and strong winds.

Meteorology websites provide generally reliable predictions for 2-4 days and they give advice on weather conditions including temperature, humidity, cloud movement, wind direction, and airspeed for most locations. In observational astronomy, strong wind can be nearly as detrimental as cloud is.

Left: **This amazing shot was taken from Penguin Air Newschopper over Phoenix, Arizona. It shows a dense, cloudburst falling out of the center of a storm cloud that has a brightly-lit, circular spreading anvil. Together, these features make it look like an atomic bomb explosion.**

Right: **The typical features of a large storm cell**

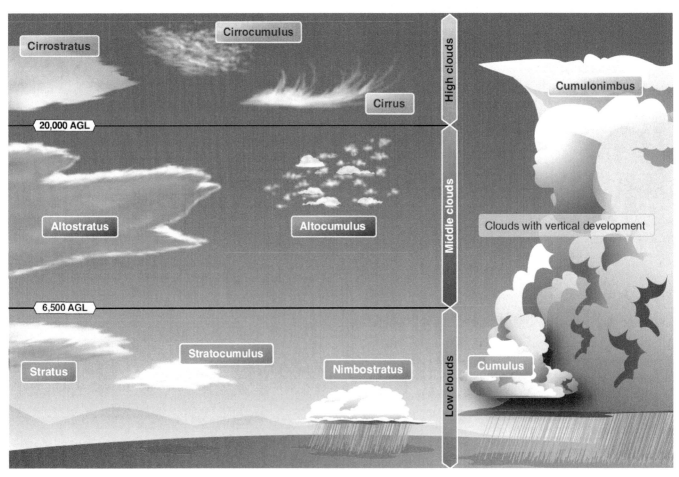

Cloud types. Credit: Pilots Handbook

WHEN DOES GOOD AND POOR SEEING OCCUR?

Clouds at different altitudes are often moving in different directions at different speeds. **Where air masses interact, this produces turbulence.**

Starlight that passes through turbulence is bent, or refracted, at different angles in fractions of a second. This makes stars twinkle to the naked eye. In a telescope, they jiggle around, and their image is blurry. Astronomers call such conditions 'poor seeing'. Under these conditions, telescopic images look similar to looking through the heat of a fire. (See page 89 of Chapter 5)

A telescope tube that has been inside a warm house, or in a warm car, will produce thermals inside the tube when it is exposed to cool air. Differences in temperature between air cells in the tube and cool outside air cause turbulence which blurs a telescopic image - just as turbulence in the atmosphere causes bad seeing. So, give your telescope time to cool down to the ambient air temperature during twilight before you start observing.

When the air is especially steady, telescopic images will have very good clarity. This is called 'good seeing'. But even with good seeing, you may have to wait a few seconds now and then for the seeing to become perfect to allow you to see very fine detail. The best seeing occurs when a large high-pressure system is directly over your observing site. At this time, there is typically steady air, with no wind, and no cloud at all levels in the atmosphere.

Temperature differences close to the ground also create turbulent air. Observing through a telescope that is looking over a warm house, a warm car, an air-conditioning outlet, a fireplace chimney, an outside log fire, or a barbecue, will suffer poor seeing due to hot air rising from these heat sources. Hot human body heat that passes through the opening of a dome causes poor seeing. Even when outside, try to avoid having your telescope pointing in the same direction that the airflow traveling past your body or others with you, does not pass in front of the open end of your telescope.

DO MOUNTAINS HAVE THE BEST OBSERVING CONDITIONS?

Contrary to popular belief, mountains do not provide good observing conditions for *amateur astronomers*. Mountains that amateur astronomers can drive up to typically have altitudes up to 2,000 ml (6,500'). They are not good sites for stargazing because they are very likely to be cold, windy, and cloudy. Even when the base of a mountain is cloud free, the mountaintop can be clouded over because mountains often create their own cloud. This happens because moist warm air from around the base of a mountain will chill and turn into mist when it passes over a mountain. I have known amateur astronomers who have built observatories on a mountain where rainforests exist only to complain about the number of nights they lose to cloud and rain! Rainforests are given their name for a very good reason.

Mountains cause poor seeing. When a relatively steady, slow-moving laminar airflow travels across a plain or the ocean, it provides good seeing because the airflow is not turbulent. But once air is forced up over the irregular terrain of a mountain or hills, it becomes quite

Air currents (winds) at different altitudes move in different directions. The three cloud layers in this image are moving in different directions causing poor seeing. The dark gray cloud is at a low level in the tropopause, whereas the orange cloud is at a mid-level. The white/light gray storm cloud's anvil is at a high level. Air turbulence occurs where layers of different temperatures and wind directions interact. Credit: Gregg Thompson

turbulent thereby causing poor seeing. Warm air mixes with cold air causing thermals that also deteriorate the seeing. When air passes over a mountain, it is forced to flow faster so this creates strong winds.

Visual observing from very high mountains over 2,500 mi (8,000') is actually *dis*advantageous for visual observing because oxygen levels are low enough to cause the human eye to be less effective at seeing faint objects. Over 3,000 mi (10,000') **altitude sickness** will affect some people who have not had time to acclimatize to such heights. People who live close to sea level are not used to low levels of oxygen at high altitudes, so they

This photograph is a very good example of a mountain creating its own cloud capping. Credit: USNO Dan & Cindy Duriscoe

Mountains create air turbulence and wind, thereby producing poor seeing, and often cloud. Credit: Gregg Thompson & Rebecca Gjerek

can become sick if they go directly to the top of a high mountain over 4,000 m (12,000'). Altitude sickness results in headaches, nausea, a loss of control over one's bowel and bladder, dizziness, and an inability to think clearly. This is especially the case when they do some exercise. Most astronomers visiting **Mauna Kea Observatory** stay at an acclimatization hotel at 3,000 mi (9,000') for a few days before working on the observatory's peak. (See Chapter 13, page 248.) This gives them time for their body to produce more red blood cells to carry more oxygen.

I can remember in my excitement when I was younger, running up three flights of stairs in an observatory on Mauna Kea only to find at the top I experienced the aforementioned effects of altitude sickness. It only lasted for a couple of minutes after I stopped, but it was unexpected and unpleasant.

A protected area on an open plain is the best place to observe from. It is far less likely to have cloud cover and much more likely to have good seeing – and it will have plenty of oxygen for good night vision.

WHY ARE MOST LARGE PROFESSIONAL OBSERVATORIES LOCATED ON HIGH MOUNTAINS?

Most major professional observatories are built on very high mountains so they are above most rain clouds and atmospheric turbulence, which is often present in the lower atmosphere. Very high mountains have very little water vapor in the air. This is important because water vapor interferes with the absorption of infrared wavelengths needed for the spectra of stars and for studying the temperatures of astronomical objects.

The ESO's 39 m (1,535") diameter **Extremely Large Telescope now under construction on Cerro Armazones peak is situated at 3,046 m (9,9990') above sea level will be the world's largest optical/infrared telescope.** Illustration by ESO.

Warning signs like this on high mountains point out that altitude sickness can cause serious health problems for those unaccustomed to doing exercise on mountains with altitudes over 2,500 m (8,000'). They also mention the risks associated with electrical storms and snow falls. This is good advice to be heeded.

This is a part of the ESO's La Silla Observatories which are perched on top of a 2,400 m mountain top in Chile's Atacama Desert. Note that the tops of the mountains are above the haze. Credit: ESO

Here we see the home to many of ESO's telescopes, including the ESO 3.6-metre telescope and the New Technology Telescope (NTT) at center at La Silla Observatory in the Atacama Desert in Chile. Credit: ESO

At the Cerro Paranal Observatory in Chile at 2,635m (8,645') has the world's largest optical/infrared telescopes seen on the mountain in the background. In the foreground is one of the ESO's Vista telescope. Credit: ESO

IS A PERFECTLY DARK SKY PITCH BLACK?

Even the darkest night sky is far from being 'pitch black'. You can easily see how dark the night sky is on a moonless night at a dark site by placing your hand, or any dark object, against it. You'll notice that the sky is a much brighter slightly bluish-gray. Trees appear black, but the sky does not. The reason the sky is not jet black is due to **airglow**.

AIRGLOW

During the day, ultraviolet light from the Sun breaks apart oxygen and nitrogen molecules in the thermosphere about 100 km above sea level. This causes ozone to be created. Ozone molecules have three oxygen atoms, whereas normal oxygen molecules have two. Charged particles in the Sun's solar wind and cosmic rays from deep space both impact ozone molecules high up in the thermosphere. This produces reactions whereby faint greenish light is emitted in the lower thermosphere and reddish-orange light in the upper thermosphere. As well, ozone molecules collide with one another and this also produces light. These faint emissions of light are seen by the human eye as the faint bluish-gray light known as airglow. It is brightest early in the night and it becomes weaker after midnight because less solar wind particles enter the air then. Because airglow light is so faint, the naked eye does not detect its color, but digital photographs do.

Being like a faint form of an aurora, the brightness of airglow varies over time and from one location to another. The more active the Sun is, the brighter airglow will be. To see excellent video photography of waves of airglow changing at the European Southern Observatory in Chile, go online to the ESO site and type in airglow.

Airglow is well seen from orbit because it is concentrated along the horizon, appearing as a thin greenish-brownish arc above the globe. If Earth had no atmosphere and therefore no airglow, then the night sky would be truly black, as is the case on the Moon. We would see numerous more very faint stars to an extra 1.5 magnitudes fainter.

Go to the end of the chapter to see a marvelous example of both waves of green and red airglow.

CAN MORE FAINT STARS BE SEEN FROM A HIGH MOUNTAIN?

Mountains offer no advantage for stargazing because we see no more stars at high altitudes. This is because, airglow exists far above our highest mountains like Mauna Kea and Mt Everest, so the sky there is no darker than at sea level. The intensity of airglow is what determines how dark the sky will be. Of course, if there is smoke, dust, or smog in the air at a low altitude site, then the air will be less transparent than on a mountain that is above such air pollution.

SKY BRIGHTNESS

Air pollution affects sky brightness. Fortunately, air quality in many western cities has greatly improved over the last three decades due to the introduction of Clear Air legislation. These laws made car manufacturers install

From space, airglow is seen as a thin greenish brown band. Credit: The International Space Station

In this picture, ninety-nine percent of all the air in the atmosphere lies at the very bottom of the thin, dark blue layer of air close to the Earth's surface. Commercial airliners fly at the top of this blue layer. Note the bright green aurora on the right. It is just above the tropopause but well under the layer of airglow. A very high red and purple aurora in the thermosphere can be seen extending well into space.

catalytic converters on cars to remove carbon particles from their exhaust gases. This legislation also made industries filter smog particles out of chimney exhausts. And residents are no longer allowed to light backyard fires. In many cities, smoky wood fireplaces are also banned.

Artificial light reflecting off smog, dust, and water vapor causes light pollution. (See Chapter 14 for more on this topic.) Unfortunately, there are still some countries in Europe that have high levels of air pollution caused by diesel fumes in the air. The burning-off of unwanted gases from oil production plants across the Middle East produces high levels of carbon particles that badly pollute the air there making observational astronomy in many regions impossible. And of course, air pollution in developed parts of China and India is particularly bad due to their coal-fired power station not having dust and ash filters like those in the most Western countries. India also has many fires from underprivileged people cooking. Cities in these areas are not places in which to do deep sky astronomy.

Yuri Beletsky photographed a brilliant display of both green and red airglow around the horizon. The Milky Way arches over ESO's Cerro Paranal Observatory. Orion is to the right, the Magellanic Clouds are at the center, and Mars shines brightly at far left.

This extraordinary photograph shows arcs of green and violet light across the sky. Both colors are airglow that occurs at different altitudes in the Earth's upper atmosphere. Credit: J C Casado

PREPARATION FOR DEEP SKY OBSERVING AT A DARK SKY SITE

For good dark-sky deep sky observing that require you to travel to a dark site, it's a good idea to plan ahead so you have plenty of time for travel to it and to set up, before night commences, especially if you are camping out. Use meteorology websites to choose cloud free times and the best observing locations.

COMFORT

It's very important to be as comfortable as possible so you can enjoy observing. If not, you will tire quickly. When you are observing with your naked eye or binoculars and looking at objects high in the sky such as satellites, a lunar eclipse, the Milky Way, or meteors, it's best to lay horizontally on a sun lounge, an air-bed, or a groundsheet with a small pillow under your head. This is a very comfortable way to observe, and because you are comfortable, you can take all the time you need to observe. Laying down stops you straining your neck and back, as occurs when looking up while sitting or standing. Also, make every effort to stay out of breezes because wind chill can make you quite cold. When you are comfortable, you will be happy to observe for hours, but when you are uncomfortable, any excuse can be enough to stop.

Food and Beverages

An adequate supply of snacks and warm beverages is a must if you're planning a long observing session. On cool nights, hot beverages such as coffee, tea, hot chocolate, milk, or soup are a good way to keep warm and to stay fresh. Keep them in a thermos flask to keep them warm. Caffeine enhances your senses and it helps stimulate you, so it will reduce your chances of becoming tired early in the evening. Tea has more caffeine than coffee. It is best not to drink alcohol or smoke tobacco while observing, as both will detrimentally affect your night vision. Alcohol can also make you drowsy. (See Chapter 7, page 117). Alcohol will increase your chances of losing your balance and having an accident in the dark. Contrary to popular belief, alcohol will drop your body temperature after an initial perception of warmth, thereby making you feel colder than you would without it.

Wind Protection

Wind chill gives the effect of the air temperature dropping by many degrees. This is because it increases

This is an inexpensive portable tarpaulin observatory protecting an 18" telescope and the observer from wind chill. Credit: Gregg Thompson

This practical, inexpensive, portable observatory for field trips utilizes 5 tarpaulins. It can be erected in ten minutes. It costs only around US$100. It provides welcomed protection from wind and stray light. Pull back roof sections tied back by elastic cord, minimize dew and protect the telescope and interior from an unexpected shower.

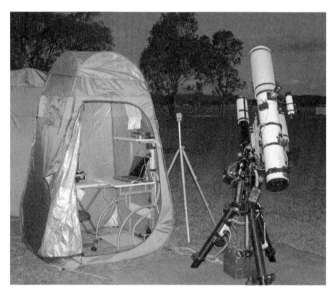

A portable observer's retreat provides welcomed protection from the elements.

A convenient pop-up tent is perfect for protecting the observer and computer equipment from cold, dew, and wind. Not shown in this photo are four steel posts that are inserted into holes in the ground around the telescope to support plastic tarpaulins that act as a windbreak for the telescope. Note the long blue dew shields on the telescope to reduce the chance of dew forming on the optics. Credit: Greg Bock

The Kendrick and Memphite tent companies have 'observatory' tents that provide a wind break for a telescope and a protected room to one end for the observer. Search for the Karl Baltz Review for more information.

Richard Bosman uses a commercial observatory tent when imaging at dark sky sites. Credit: Richard Bosman

evaporation of body moisture and this makes you feel much colder than when you are out of the airflow. Prolonged exposure to even a light breeze can be very uncomfortable. A breeze can also cause your telescope to vibrate. To avoid wind, set up behind an existing windbreak like a building of a tall hedge, or make your own portable observatory with tarpaulins supported by tent poles and ropes. Have the opening on the downwind side. Have the walls about the height of the telescope or higher.

If you are doing astrophotography, you can often sit relatively motionless for hours, so you can become very cold in winter. Being able to retire to the warmth of a small commercially available tent can be a godsend. A warm, dry tent environment is needed for using a laptop while the telescope is photographing outside.

Dressing for the Occasion

As we have discussed previously, on average, nighttime temperatures are around 10 degrees colder than during the day and clear nights are the coldest. At remote

Left: **Freezer suits or ski clothes are great for minimizing cold. Wear a beanie or balaclava under the hood. Snow boots and two pairs of socks and gloves are essential. The inner gloves can have the fingertips cut out so you can handle small objects when you briefly remove your thick outer gloves.**

Right: **Here we see Christian Viladrich in France suited up astrophotography in for extreme cold where the ground is covered in ice.**

locations at altitude, it can become quite cold, even in summer, so be prepared. Of course, in summer in tropical latitudes near a warm ocean, you can often wear light clothing and be comfortable, but even in these locations, it can become quite cool as the night progresses, especially if a sea breeze picks up.

Inexperienced observers often underestimate how cold it can get when you are standing relatively still for long periods. When I was a novice, experienced observers taught me a good rule for observing away from home - always take twice as much clothing as you think you will need. So often, you end up needing it all when it turns colder than expected. It's also good to take a change of clothes in the event that you get wet for some reason like slipping over on wet ground, spilling a drink in the dark, or getting wet from an unexpected rain shower.

It's best to have clothes that allow you to add more layers when it becomes colder. If it gets warmer, you can peel them off. In winter, be sure to keep your head covered because a lot of body heat is lost from the head due to the brain being the organ that requires the most blood.

On some winter nights, it was so cold that my eye sometimes froze onto the eyepiece! The first time this happened, I was shocked. I was frightened that if

I pulled my eye away I would tear off the outer cells of my cornea! I had no warm water within reach so after a few seconds I realized that I had no option but to pull away as the freezing would become worse. As I pulled away, I heard the cracking sound of ice breaking. To my relief, no damage was done. It was only the surface water on my eye that had frozen, and not my eye. Had I left it much longer, I may have had a more serious problem.

On another occasion, it snowed briefly off and on during a night's observing run. A colleague and I were using a 60cm (24") telescope in an observatory at Siding Spring Observatory that thankfully provided welcomed protection from a bitterly cold stiff breeze outside. We had to make observations directly downwind. To keep warm and alert, we had consumed a few cups of hot tea and coffee. When the wind dropped a little, we stepped outside to relieve our bladders. To our surprise, above the low howling of the wind, we could hear a crackling sound like shards of glass breaking on the rocks below our feet. Using a torch to see what was causing this noise, we were amazed to see our steaming urine stream was quickly turning into ice crystals before it hit the ground!

If you go observing in very cold locations, you may be at risk of being snowed in by an unexpected snowstorm, so be sure to take extra warm clothing, warm fluids and a fully charged mobile phone in case of an emergency.

Amateur astronomers devise many nifty things to make observing more comfortable and more practical. Here are some ideas that I developed that are simple and effective.

Managing Dew

Dew can be an annoying problem when you are observing outside when the humidity is high and the temperature drops. Many astronomers have had dew damage their reference books and star charts etc. To overcome dew, convert an existing lightweight cardboard box to become a dew shield. Cover it with Contact plastic film to stop it getting wet. Alternatively, build a lightweight ply box. Make the box big enough for you to put your books, charts, or a laptop inside it. Have it high enough to allow you turn pages easily and so you can write notes easily inside it. To stop dew, install a low wattage incandescent torch light bulb on a dimmer inside the top to provide dim light but also just enough heat to stop the inside from dewing. It can be powered by a battery. Incandescent bulbs produce heat, LEDs do not. You will need the dimmer to run the light at a low brightness. Place a light shield over the front of the light so that direct light cannot enter your eyes. The front flap can be dropped to stop breeze blowing papers around when the box is not in use. Closing it will keep light out of your eyes when you are not using it, and this will keep the interior warm. The box will need to be weighted down or held in place by bungee cords to stop wind gusts blowing it over.

A Portable Telescope Table

A table at the telescope is a big advantage. A lightweight, easy-to-carry, folding picnic table or a card table is a good choice. Because Newtonian telescopes have an eyepiece that sits high, it's a good idea to have a high table so you do not have to keep bending down. A simple solution is to raise the table by adding extension legs. I used square sections of aluminum tubing that fit snugly over the existing leg. An extension leg fits into the other end of the joiner. The extension legs are taken off for transportation. See illustration above.

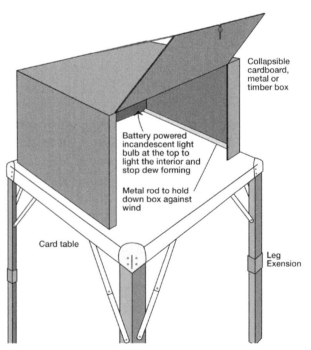

A dew box protects books and star charts from getting wet.
Illustration by Gregg Thompson and Nicole Brooke

Protection from Rain

Sometimes there will be occasions when, out of nowhere, a rain cloud passes over and starts to rain. It's a good idea

Waterproof zip-off covers for 200 mm and 460 mm telescopes

Telescope covers are quick and easy to use, and they allow telescopes to be left outdoors ready for use. Castor wheels on the Dobsonian allow it to be easily rolled around and placed under cover when not in use. Credit: Gregg Thompson

to always have a raincoat handy for such occasions and it's also advisable to have a tarpaulin handy to throw over your telescope and another for your portable work table so your books, charts, and eyepieces don't get wet. As wind often accompanies a rain shower, be sure to have a means of holding your tarpaulin in place with weights attached to the edges. Or use elastic bungee cords.

To save the time of setting up and packing away your telescope for each observing session, you can have an upholsterer make up a zip-up, waterproof vinyl cover for your telescope so it can be left in place in the backyard, or on a deck. These covers are especially practical and effective. The slip-off cover can be quickly removed for observing and quickly replaced at the end of the evening or during a shower. Commercial covers that are advertised in astronomy magazines can be purchased online. It's important to have a cord that pulls tight around the telescope's base, so wind cannot blow dust into the telescope when it is not in use. Dust caps should be placed over both ends of the telescope's tube before putting the cover on.

AN EYEPIECE CASE

To stop eyepieces dewing up in winter, you can keep them in a dust-free pocket in your apparel so that your

A foam-lined camera case is excellent for storing eyepieces, filters, and tools. Credit: Gregg Thompson

body heat can keep them warm. Another option is to have some heated stones or thick metal discs placed in your eyepiece box. Alternatively, have your eyepiece box heated by passing electricity from a battery through an insulated resistance wire that acts as a mild heater running under your eyepieces.

It's best to keep all your eyepieces in a hard camera lens case so they are all together and protected from unplanned impacts. You can also use the box to hold some tools like a spanner, a screwdriver and Allen keys etc. You can also include a torch, insect repellent, batteries, and filters. It's a good idea to wrap plastic film-coated foam around your eyepieces to help protect them if they are accidentally dropped. This can also act as a means of helping to block stray light from entering your eye by pressing your eye into the foam surround.

In winter, the thick foam insulates the cold metal from touching your eye and hands if your eyepieces are not heated.

THE EFFECTS OF BODY HEAT

It is important to mention the detrimental effects that body heat can have on a telescopic image. When using a telescope, avoid having people stand in front of your telescope tube, or in an airflow that will travel in front of the telescope. Body heat drifting in front of incoming starlight will blur the telescopic image quite seriously.

To see how much heat comes off your hand, place it in front of your telescope tube while it is pointed at a bright star or the Moon. Remove the eyepiece and look through the eyepiece holder to see the heat waves drifting off the silhouette of your hand. It looks like you are on fire! After seeing this amount of heat from just one hand, you might imagine how much heat rises from a person's whole body, let alone from a group of people standing near the telescope.

In the next chapter, we will learn how our eyes work so we can use them to their best advantage to observe maximum detail and very faint light.

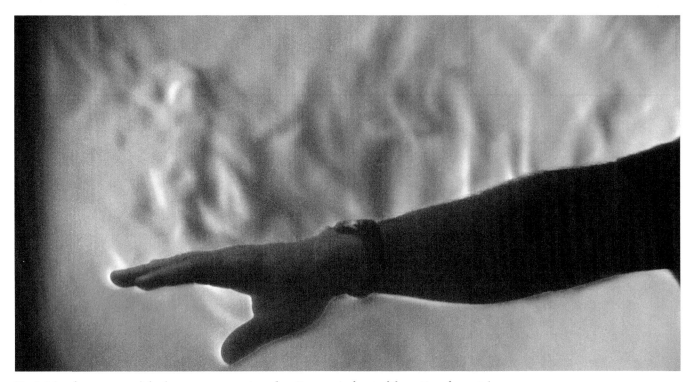

Heat rising from a person's body causes poor seeing when it passes in front of the optics of your telescope.

Chapter 7
MAKING THE MOST OF YOUR EYES

To be a good observer not only requires a good instrument and good seeing conditions but also knowledge of how to use your eyes to see maximum detail and very faint light.

In this chapter, we will look at:

- the value of visual observations compared to photography,
- why it's important to take your time when observing,
- how our eyes work with regard to astronomy,
- how you enhance our vision and,
- how you can test our eyesight.

THE ADVANTAGES OF VISUAL OBSERVATIONS

When you are seeing an object in a telescope directly with your eyes, you are seeing the real object, and it feels like it is there in three dimensions. Whereas when viewing photographs, they do not have the same feeling of being the real thing. And they are clearly two dimensional and limited in size. When seeing an object in a wide-field eyepiece, you feel like the edges of the field are so wide that it's like you are looking through a wide open portal into space. Another advantage of a visual observation over photography is the ability to see far more subtle tones and ranges in brightness than photographs can record. Visually, even the brightest stars are pinpoints of light, but in photographs, the brighter a star is, the larger is its diameter, so they don't look real like they do in a telescope.

Ⓒ A classic example of how powerful a visual observation can be compared to a photograph occurs when people are shown an impressive Hubble Space Telescope picture of Saturn in rich color. Surprisingly, most people typically show mild interest, but when they look through a telescope at Saturn's globe surrounded by its rings and its moons suspended in space in apparent three dimensions, they are invariably excited to see this. This real gives them a 'Wow!' reaction, yet the telescopic image of Saturn is nowhere near as large, or as colorful, or as detailed as a Hubble Space Telescope photograph. Surprisingly, yet it wows people far more because most people have never seen Saturn for real. The same 'Wow!' experience also occurs when people first see the Moon's stunning, sharp, high contrast detail, or when they look into the innumerable number of stars in a large globular star cluster in a wide field eyepiece in a dark sky. Visual observations of a large bright open star cluster with colored stars can also be more impressive in a good telescope than a picture is. But for faint objects like nebulas and galaxies, a photograph will provide a much more detailed image with lots of colors that the eye cannot see. **Both visual observations and photographs have their advantages, so they should be used to complement one another.**

HOW TO ENHANCE YOUR OBSERVING SESSIONS

Ⓒ To maximize the joy of visual observing, before you start, look for photos of the object(s) that you want to observe. You should be able to find them in *STARGAZING*, or on the Internet. These photographs will give you a good idea of the main features you may see in the brightest regions of the object. But because most photographs are taken with large apertures that can capture fine detail and faint regions, it is understandable that you will not see visually the detail that photographs capture. However, in many photographs, the bright portions of an object may be burnt out, but visually, you have a better chance of detecting subtle details in these regions. It's also a good idea to read about the latest discoveries related to the objects that you plan to observe. This knowledge gives you a greater appreciation of what you will be looking at, so this will make your observing more exciting.

PERSEVERANCE PAYS BIG DIVIDENDS

When people first start observing, they tend to look at objects rather quickly. In doing so, they only see a small fraction of the detail that is there. **It's important to take your time when you look into the eyepiece of a telescope so that you give yourself plenty of time to inspect the object and its field closely.** The atmosphere is often unsteady, so you usually have to wait for those moments when parts of the image become sharp - so don't rush. Sometimes, one part of the image will briefly become very clear, and then another part soon after that. To see all parts in maximum clarity, it may take you some time if the seeing isn't excellent. This is true for the discs of Mars, Jupiter, and Saturn when observing at medium to high power.

It's best to sit on a chair or stool so you can gaze into the eyepiece in comfort for as long as you need to. By taking your time, your memory has a chance to build up an overall image of the object by putting together glimpses of fine detail. You end up with a 'mind' picture that is far more detailed than what you can see when looking at an object briefly. **Perseverance delivers the best results.** To fully maximize the detail you can see, it's a good idea to draw the planets or deep sky objects because this forces you to check the detail carefully to get it right. (See Chapter 10.) And it gives you a permanent record of your observation.

If objects are very faint, you will need to use averted vision. (See page 117.) This also takes the eye and brain time to build up a complete image. With averted vision, you have to inspect each faint part of the field bit by bit. Once you practice taking your time to look for fine detail, you will do this intuitively.

HOW OUR EYES WORK

If you are to get the most out of your eyes when observing, it's important to know how your vision works – just as you need to know how a camera works to get the most out of using it. The following sections explain various aspects of your vision that have the most impact on stargazing.

Light first passes through your eye's cornea, the clear film over the outside of your eye. It then passes through the **pupil**, the colored muscles that expands and contracts to control the amount of light lets in. After this, light then through your eye's **lens** to bring it to a focus at the back of the eye. But before it gets there, it passes through the clear **vitreous humor** jelly in the eye before it finally reaches the retina cells on the back of your eyeball. The retina cells pre-analyze the light information before sending it on to the brain for further evaluation.

The black pupil is the eye's window into the outside world. In bright daylight, our colorful iris muscles surrounding the pupil close down our pupil to its minimum diameter, which is just 2 mm. When it is dark, it opens up to a maximum diameter of 7 mm to let in the maximum amount of light.

There is less of a chance of there being a flaw in the cornea that covers the small 2 mm area at its center than there is across the entire 7 mm area. This is why our vision is clearer in daylight than under low light. When the pupil is closed to its smallest size, we have the greatest depth of field – meaning that we can see things that are both fairly close and also those that are far away in clear focus.

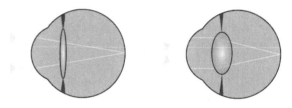

The eye's lens is clear and flexible. It has muscles the whole way around it (shown in red). They stretch the lens out to become thin, so it can focus on distant objects (left), and they compress it to make it thick to focus on nearby objects (right).

HOW OUR EYES FOCUS

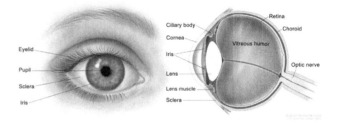

This diagram shows the eye's main parts.

Note that the cornea is the clear region that covers the front of the eye. Wrinkles in this will cause astigmatism, and scratches will cause flares when looking at bright stars, the Moon, and bright lights. Clear, jelly-like fluid fills the vitreous humor behind the lens (red). In front of the lens is the aqueous humor (blue). Credit: Terese Winslow

TESTING YOUR VISION

To make scientifically valuable visual observations, it's necessary to know the state of your vision. This should be recorded on your observation forms. This information is useful for comparing your eyesight to that of other observers, and to see how it performs as you age.

There have been times where I have seen something quite easily through a telescope, yet another observer using the same telescope has said that they could not see it, because it was too faint or, the detail was too fine.

This can be caused by differences between:

1. Observers' Observational Experience, which can only be gained over time with practice.
2. Observers' Visual Acuity – this refers to how much detail and observer can see.
3. Observers' Threshold Vision – this refers to the faintest object an observer can see.
4. A combination of these things.

A person who has good visual acuity may not have good threshold vision, or vice versa. These two forms of vision are different and not linked.

SEEING FINE DETAIL

The finest detail a nearly perfect human eye can see is 1 minute of arc (1') or, 60 seconds of arc (60"). The degree of fine detail that we can see is related to **1.** How well our eyes can focus, and **2.** the imperfections that exist in one's cornea –the clear outer covering over the iris and the pupil.

At the back of the eye in the **retina** is a small area known as the **center of vision**. This is where we see by far the most detail. However, it is an extremely small area. Incredibly, the very fine detail we see here is only the size of two letters in the type on this page when it's held at 400 mm (16")! To test this, stare at the cross in the middle of the jumbled small type following:

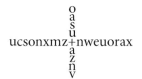

Without moving your eyes, see if you can tell what the letters are beyond the first letter on either side of the cross. You should only be able to see one letter if you do not cheat. (We automatically move our eyes, so it takes some practice not to cheat.) Incredibly, just the area of the cross is how tiny our center of high clarity vision is!

VISUAL ACUITY TEST

A person's **visual acuity** refers to the degree of detail they can see in either eye.

The test below will determine your visual acuity in daylight and also under the low light of full moonlight. When you perform these tests, record the numbers. Keep a record of these numbers so you can repeat these tests every few years in order to track changes in your eyesight over your lifetime. If you wear glasses for seeing into the distance, keep them on. It is also interesting to do these tests with children and friends of different ages to see how much aging affects how much detail we can see. Often one eye can be better than the other. Each eye should be tested separately. To do this, test one eye by holding your hand in front of other but do not have your hand touch your eyeball. This way you can keep both eyes open.

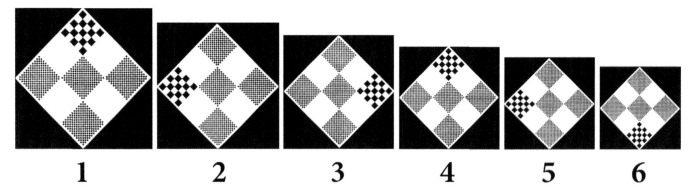

Perform this test in full sunlight. Print out this test page and place it at eye height at exactly 3.3 m (10') away. Look at each of the squares above and record the number of the smallest square in which you can *definitely* see the checkered pattern.

You will only fool yourself by guessing so record the number where you are sure you can see the checks. It's best to have someone beside the chart to check whether you are right. The number where you can see the checkered pattern becomes your visual acuity number for bright light. Test both eyes. If you prefer one eye to look through the telescope (most people use their right eye) then record the number for that eye. However, if your other eye is better, then you should observe with that eye.

The Full Sunlight Visual Acuity Test Chart

When recording your visual acuity figures, you should record your score for the daylight test first, and later test it by the light of a Full Moon. E.g. 4/2. Do not look at the Moon in the minutes before this test. The Full Moon test is the most important for visual astronomy, as your eyes are typically dark adapted - unless you are observing the Moon in which case, the Daylight test would be most applicable.

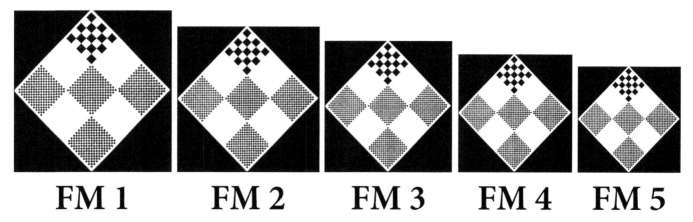

Do this test with it illuminated by the light of a Full Moon and no artificial light. As with the previous test, stand exactly 3.3 m (10') from the test. Ensure that your pupil is fully dilated by being in the dark for a few minutes. Do not look at the Moon in the minutes before you do the test. Place this test sheet against a darkish background. Ensure that there is no artificial light illuminating it. Test each eye separately to see which eye is best to observe with. As with the previous test, the number of the smallest square in which you can see the checkered pattern becomes your visual acuity number for full moonlight (i.e. low lighting conditions).

OBSERVING FAINT OBJECTS

The Horsehead nebula (center) and the Flame nebula (left) are both very faint nebulas. To have a chance of seeing them in a dark sky with a large aperture telescope, your eyes must be fully dark adapted, and you must have experience in using averted vision. Credit: Simon Addis

Dark Adaption

It is very important for your eyes to be fully **dark adapted to see faint stars, deep sky objects, and comets.** It is not necessary for observing bright objects like the Moon and the planets.

For most people with good vision, the eye takes around 6 minutes to become *almost* fully dark adapted for general observing, and a little longer for people over 70. However, after two and half minutes our eyes are *fairly* well dark adapted, but not fully. **For your eyes to have maximum sensitivity to very faint light takes 20 minutes in total darkness.** This is impractical and for most observing it is not necessary. It is only required to see extremely faint objects like very distant galaxies in clusters or the faint extremities of nebulas.

To achieve maximum dark adaption, there can be no stray light whatsoever – not even airglow entering your eyes. Your eyes must not be able to see *any* light other than what is in the eyepiece. This will require a dark cloth over your head. A much better option is a **light shield around the eyepiece** shown on page 117. It also helps a lot to reduce airglow by observing from inside a dark observatory with no lights on and only the minimum amount of skylight entering the slit of a dome or a dual roll-off roof observatory. Observing inside a dark observatory through a slit will make your eyes much more dark-adapted than observing outside under the night sky where airglow covers the whole sky, thereby considerably reducing your sensitivity to very faint light. You might assume these measures are over-the-top, but you will find that they make a surprising difference to seeing very faint objects. Most amateurs have never thought to try this, so they are unaware of how effective it is. It surprised me when I tested it.

To maintain good dark adaption when observing at home, try not to go into brightly lit rooms for a drink or other things. If you do need to go inside, wear dark red goggles or sunglasses and leave bright lights turned off inside. Just have a soft lamp on. Do not look directly at any light. Rather than going inside for a drink or food, take what you need with you at the outset, so you don't have to enter bright light and have to dark adapt afterward.

Substances That Reduce Dark Adaption

Alcohol has a detrimental effect on dark adaption, as do diseases of the liver, diabetes, and low levels of oxygen in the blood. Drugs that contain **Benzedrine** will affect your night vision. Tests have shown that the amount of **nicotine** in just one cigarette smoked as much as half an hour before observing, will noticeably reduce one's night vision.

By contrast, clinical tests show **that comfort and freedom from anxiety maximize the ability to dark adapt.**

Test the Power of Full Dark Adaption

Ⓒ As a graphic demonstration of the power of full dark adaption, spend at least 5 – 10 minutes in a *completely* dark room in the country under a dark sky (maybe chatting or listening to music) to allow your eyes to become totally dark adapted. Then go outside with your hands over your eyes and then quickly look at the sky and the Milky Way under a naturally dark sky with no stray light. You will be amazed at how brilliant the Milky Way and the sky appears when you first see it before your eyes lose their maximum dark adaption over the following several seconds. This is why you should try to get fully dark adapted when looking for very faint objects. It makes a major difference, especially when you are using a Deep Sky filter on nebulae because the filter will darken the background sky helping to maintain your maximum dark adaption. When looking for faint galaxies, you can't use a filter, so use high magnification to darken the background.

The Ability of the Eye to Be Fully Dark Adapted

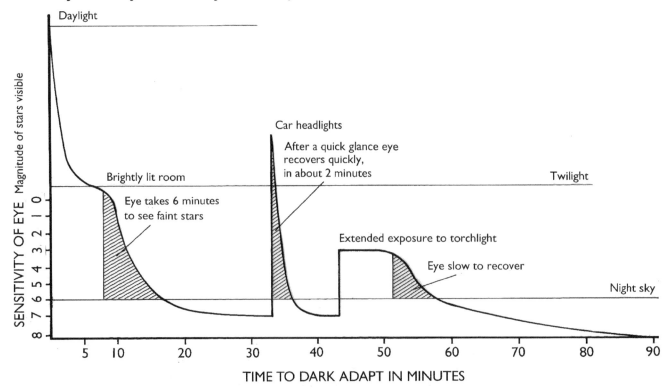

This graph shows how a dark-adapted eye can recover fairly quickly within a couple of minutes after glancing at very bright car headlights but surprisingly, extended exposure to even relatively dim torchlight or airglow can take a few minutes to become almost fully dark adapted.

Airglow is seen here as the greenish atmospheric layer above the Earth. This photograph was taken from the International Space Station when it was passing over the central east coast of Australia. Brisbane and the Gold Coast are the strip of adjoining large city centers. Note the fainter and broader, red airglow higher up. Credit: NASA ISS

Even the keenest-sighted observers cannot see stars much fainter than magnitude 6 because the background airglow is around magnitude 6.2. Stars fainter than magnitude 6 are lost in the brightness of the airglow. If the sky was totally black, as it is on the Moon, then we would probably see stars as faint as magnitude 7.

This amazing image taken by Yuri Beletsky at the ESO shows the green light of airglow near the center coming from around 100 km above the surface. To the right, the red light of airglow comes from higher in the atmosphere. Orion lies above the horizon on the right, while the center of the galaxy is setting on the left. The Clouds of Magellan are at center.

Night Blindness

An inability to dark adapt will occur if one has a total lack of **vitamin** A (retinol) in one's diet over many months. Vitamin A deficiency is very rare because the liver typically stores enough to last for months. The body only needs a relatively small amount of this. It is available in many foods such as fish oil and dairy products. Eating many carrots to enhance your night vision is a myth.

Night blindness can occur early in life from several diseases. It makes driving at night very difficult. It's like driving at night with dark sunglasses on – a very scary experience if you try it - but as a passenger!

Night blindness can also be caused by an injury to the eye. Those with night blindness are obviously not suited to deep sky observing.

Torchlight

To maintain dark adaption as much as possible, you will need to use a dull red light that is just bright enough to allow you to see your star charts, reference material, or drawing paper if you are sketching. Our dark-adapted eyes are least affected by faint red light, so use a dimmable red LED. You can use a normal torch and cover it with a few layers of red cellophane, or even white paper, which gives a soft, dull brown light. If you are using an electronic display device such as a mobile phone, then dim the display as much as possible.

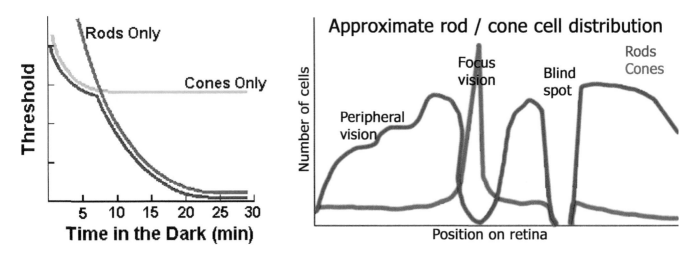

Left: Note how the cones cells in the eye that show the most clarity, only dark adapt to a minimal level to about 4th magnitude. The rods cells which see faint light, but much less detail, take 10 minutes to see 6th magnitude stars and 20 minutes to become fully dark adapted to see below 7th magnitude - against a black background.

Right: This graph reveals the density of rods and cones across the retina. Note that cones (red) have the most density in a very small area at the center of vision. They quickly reduce in number beyond here. The density of rods cells (blue) also fall off the further they are away from the center of the eye. Note that there are no rod cells at the center of vision because cones saturate this small area. Both rods and cones do not exist at the blind spot because this is where the optic nerve that runs to the brain leaves the eye.

Stray Light

When your eyes are dark adapted, even the faintest artificial light seems bright when it enters your eye, even obliquely. Such light could be a very distant porch light on a farmhouse over a kilometer away, or some distant car headlights. To maintain your night vision, you need to block out stray light by placing an object between it and yourself - such as a car, a tree, a hedge, or a tent etc. Alternatively, erect a tarpaulin as a light shield around your telescope. Another simple solution is to place a dark cloth over your head but this makes it easy for the eyepiece to fog up and it can be hot in summer. The best solution is to make light hood to surround the eyepiece. This will eliminate airglow and allow you to have full dark adaption, so you will be much more likely to see very faint objects. Ⓒ This can be made from light plastic or cardboard pieces taped together and covered in a waterproof Contact film to repel dew. It should extend to your ears so you cannot see beyond its ends.

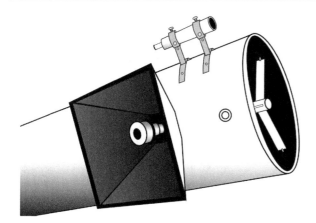

This simple lightweight eyepiece light shield eliminates all stray light when searching for very faint objects.

The shield fits snugly over the eyepiece drawtube. The shield must be large enough to allow the observer's head to fit inside it when looking into the eyepiece. The inside must be matt black so that it reflects no light. The shield needs to be light-weight, so it does not overbalance the telescope.

AVERTED VISION

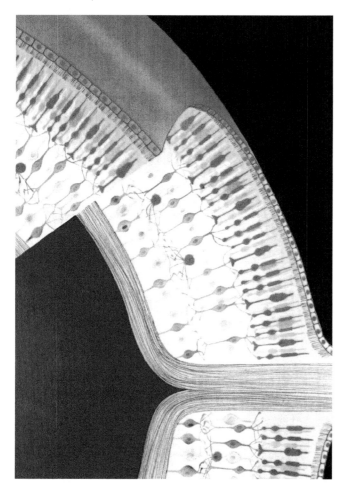

This diagram shows the eye's cone cells (red) and rod cells (yellow) between them. These cells cover the inside of the back of the retina. The lilac fibers are nerves at the front of the retina that run to the optic nerve. They are transparent and/or very thin so that they do not block light reaching the cones and rods.

It seems like the eye has been designed back to front because the light detecting cells are at the rear of the retina, not at the front. Regardless, this design works very well. The diagram on the following page shows the construction of rods and cone cells and the choroid cell layer that lines the very back of the retina to absorb light and stop it reflecting around and blurring one's vision. These cells reflect red light making people's eyes appear red during flash photography.

When trying to see faint objects with the naked eye or in a telescope, the use of averted vision assists greatly. It allows you to easily see faint objects that are invisible with direct vision.

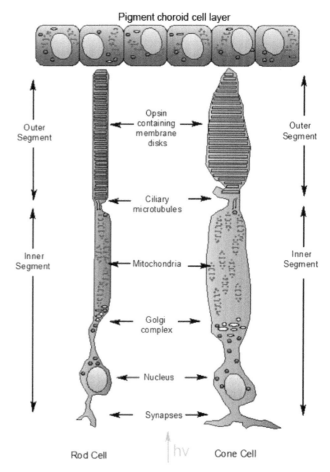

Here we see the difference between the construction of rod and cone cells. The yellow arrow at the bottom indicates the direction from which light enters the eye and these cells.

Rod cells see only in black and white and they do not show much detail, but they can detect much lower levels of light than cone cells can. They are also good at detecting movement.

Ⓒ To use averted vision, you must concentrate on faint light coming from between 15° to 30° away from your center of vision. **Rod cells** in your retina that detect faint light, dominate here. They will not show detail but they will detect very faint light. To maximize averted vision, flick your center of vision around your field of view every second or two while concentrating on the area of interest. This allows faint light to fall onto fresh, unsensitized rod cells.

Cone cells detect fine detail and color. There are three parts to cone cells; one part for red, blue, and green light.

THRESHOLD VISION TEST

Ⓒ **The threshold vision test determines how sensitive your rods cells are to faint light. When a healthy eye is fully dark adapted, it is 100,000 times more sensitive to light than when it is adapted to bright daylight!**

To test how well your eyes can see faint light, use the **Threshold Vision Test** on page 119 following. This test determines the faintest shape that you can see under a dark, moonless sky when your eyes are fully dark adapted. To do this test, use a good quality laminated copy of the test sheet provided. Place it at your feet while you are standing. Your eye should be about 1.5 m (60") above the test sheet. Allow only sky glow to illuminate the test. **This test must be done under a dark sky** where there is no stray light falling on the test, or entering your eyes. If there is a little stay light, then place the test sheet inside a dark cardboard box and keep any the stray light out of your eyes because it will diminish your ability to see the faintest shapes. Be sure you have allowed at least a full 5 minutes or more in a very dark environment and not looking at the sky for your eyes to be well dark adapted before doing the test. Test each eye separately to see which is best to see faint light. Note the number of the faintest shape you can see on the test sheet. To be sure, describe the shape to someone assisting you who can look at it under soft light. Do not try to remember the shapes. Drop the sheet so that you do not know which way it landed. To eliminate cheating you can cut them out and have someone place them randomly on a black background like black cloth. Your score is the faintest shape that you can detect with each eye. On your observation forms, use the number for the eye you will observe with. (For Observation Forms, see Chapter 10, page 176.)

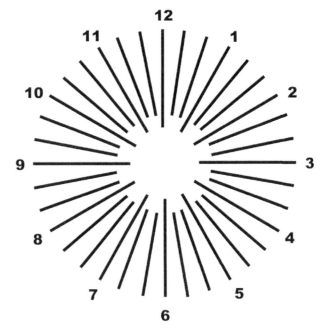

If all these lines are not equally wide and equally dark, then you have astigmatism. Test each eye separately by placing your hand over one eye and then the other.

The degree of difference in the thickness of the lines will determine your degree of astigmatism. Rotate your head to see if the thicknesses of any lines change position with the change in the angle of your head. If they do, then this proves you have astigmatism. In a good eye, all the lines will be the same width regardless of whether you rotate your head or not.

ASTIGMATISM TEST

The outer surface of the eye is called the cornea. As our eyes age, like our skin, we usually develop ripples or wrinkles in the cornea. Parallel ripples cause astigmatism. You can see a similar effect when looking through old, cheap window glass made in the 1950s or earlier. Poles or tree trunks appear to expand and contract as you move your head past waves in the glass. Microfine, dimples or wrinkles in the cornea cause double or multiple images of the Moon or bright stars. Fine scratch wounds in our corneas cause bright objects to have spikes. Astigmatism can be largely overcome by correctly prescribed glasses.

The Threshold Vision Test

Each figure from 1 through to 9 is 5% lighter than the previous one. 1 is 95% black and 9 is 55%.

FLOATERS

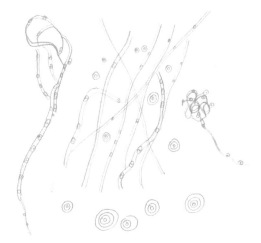

Floaters in the eye are typically curly, transparent, string-like chains of foreign cells.

You have probably seen semi-transparent, grayish wriggly floaters when you look at a bland surface like a clear blue sky, your monitor, or a bright plain white wall. They are not noticeable most of the time because we are moving our eyes around. Floaters are not age-related because everyone has them, even children. They are foreign cells that float around in the vitreous humor, the viscous fluid that fills the ball of the eye. They look like chains of cells. Floaters that are close to the retina appear the most clearly defined. Those further away are out of focus, so they look like blurry curly hairs. When they overlap one another, they appear like small dark gray hazes. They can be annoying when they float in front of a planet or bright objects when you are looking for fine detail. They drift downward across your center of vision. But when you move your eye around, this causes the fluid in your eye to swirl a little, so this makes them 'roll' up only to drift back down across your center of vision some seconds later. At present, there is nothing that can be done about annoying floaters, but in the near future, the vitreous humor in our eyes will be able to be replaced with a perfectly clear fluid.

AS EYES AGE

When I was a child and into my teens, I was able to look at flowers and insects in sharp focus even when they were as close as only 75 mm (3") from my eyes. I could see insect's compound eyes, their hairy legs, their scales, and their scary mouths as if I was using a strong magnifying glass! Yet when I looked away to infinity, my young eyes refocused immediately to give me a clear image of distant objects. Very few children realize they have this exceptional power of microscopic vision to see things in such fine detail when looking at things very closely. Unfortunately, the ability to have this large range of focus diminishes with age.

As cells in the lens grow old over time they lose their elasticity and therefore their ability to change shape to focus on nearby objects. This is known as **presbyopia**. Generally speaking, in our mid-20's to mid-30's, the closest most people can focus on an object is about 200 mm (8") away. This distance gradually lengthens to around 300 mm (12") by our mid-40's. This is when one finds the need for reading glasses, particularly when in low light because that is when our pupil is opened up to its widest, so this makes defects in the cornea more apparent. In our mid-40s, the distance for a clear focus becomes longer than one's arm! This is when reading glasses are required. When one reaches their 50s, it becomes increasingly difficult to change focus between the stars at infinity and a star chart that is 300 mm (12") away without using reading glasses. Those with normal vision become more longsighted with age i.e. they can see things in the distance clearer than things that are close.

If you wear **glasses for long or short-sightedness,** you do not need to use them when looking through binoculars or a telescope because the instrument's focuser will bring the object into a clear focus without you needing to wear glasses. The benefit of not wearing glasses when looking through a telescope is that you can place your eye close to the eyepiece and thereby see a much wider field of view. Binoculars allow for the focus to be adjusted separately in both eyes because many people have eyes that have a different focus. (See Chapter 8.) If you have acute astigmatism, you will need to leave your glasses on when observing.

Aging can also reduce the elasticity of the iris muscle. This causes the eye's pupil not to be able to fully open in darkness, thereby making faint objects impossible to see. However, this usually does not occur until after 70 or 80 years of age.

After 75, the blue end of the spectrum gradually becomes less visible because our lenses become more yellowish due to the accumulation of old cells.

These old cells act as a yellow filter that attenuates blue light - just as brown or yellow sunglasses do. By our late 70's, deep purples typically become less noticeable. The brain compensates for this to some extent by adjusting the eye's color balance, so we are not so aware of the loss of the far blue end of the spectrum. Note how blue-green Claude Monet's paintings of waterlilies were in this youth and how much more red to yellow-green they became in his late years.

Cataracts

People who have been outdoors in sunlight much of their life often develop cataracts in later life due to excess UV light. Some cataracts are genetic. **Cataracts cloud the inside of the lens thereby blurring the transmission of light.** Left untreated, this causes blindness. Today, thanks to bionics and surgical technology, cataracts can be removed by replacing the damaged lens with an artificial one.

Visual astronomers, who have had a **lens replacement** in their later life, often comment that they are surprised at how much better they can see the colors of violet, blue, and turquoise – but only for the first few days after surgery. This is not due to the replacement not being effective but rather, remarkably, the brain adjusts to this improved color vision so fast that the difference is no longer *consciously* noticed. But blue-green planetary nebulas and comets return to how colorful they were in one's youth. The reason reds look more intense is because the purple-blue end of the spectrum adds to them.

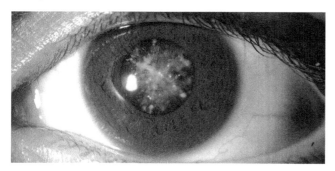

White cataracts in the lens gradually grow worse and cause blindness.

Macular Degeneration

If you are over 60 and you notice blurred or darker vision at the center of your vision, then you could have symptoms of **age-related macular degeneration**. This causes straight lines, newsprint, or faces to appear wavy because the surface of the retina is detaching from its backing. Colors become less vibrant and it is common to require more light than usual to see. In time, the whole center of vision can disappear making it impossible to see any detail. If you have any of these symptoms, you should see an optometrist promptly to have your eyes tested. Timely laser treatment can sometimes slow further loss of sight.

THE IRIS

The eye's iris is the colored streaky muscle area surrounding the black **pupil**. It is so distinctive to each individual that it can be used like a fingerprint for identification purposes. It never changes after 2 years of age, unless the eye is damaged. This fact makes the basis of the old alternative 'health' fad of Iridology untenable.

Over 70 years, the iris becomes less elastic, so it cannot open the pupil as wide as it could in younger years. This causes night blindness.

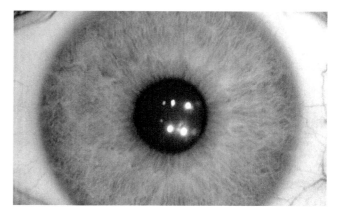

Here we see the colored muscle fibers of the iris. These small muscles automatically adjust the diameter of the pupil to control the amount of light that enters the eye. In bright light, the pupil's diameter is only 2 mm (¹⁄₁₆") in diameter but in dim light, it expands to 7 mm (¼") in diameter to let in as much light as possible.

Knowing how to make the most of your eyes and your telescope is the secret to becoming a good observer.

Chapter 8

TIPS ON OBSERVING WITH THE NAKED EYE & BINOCULARS

NAKED EYE OBSERVING

There are so many things to see with the naked eye when stargazing that it can keep you occupied and enthralled for a considerable time. Chapter 4 highlights the variety of astronomical objects and phenomena that you can observe without optical aid. In Volume 2, you'll discover that you can watch the Sun and the Moon move around the sky and you'll learn why they do so. With your naked eye, you'll also learn to identify the planets and watch them continually moving through the constellations and changing in brightness. It's also fascinating to discover that stars have different colors, and what their colors mean.

In Volume 3 Chapter 8, you will learn about the brightest stars and the major constellations, and you'll understand why the constellations change from one season to the next.

Naked eye stargazing gives you a very good appreciation of where the bright deep sky objects lie amongst the constellations. It's good to have that knowledge before purchasing binoculars, or a telescope. You'll notice that each night the stars move across the sky in arcs that have different curvatures depending on where they are located in the sky. The naked eye is best for observing the Milky Way's star clouds and its dust lanes.

Our naked eye is superb for looking at great comets that have tails that can be tens of degrees long. In 1910, and during 1986, Halley's Comet was only able to be seen in its totality with the naked eye because the tail extended across much of the sky! This was also the case for all the other great comets. (See Volume 2 Chapter 12, page 480.) The Earth's shadow in the atmosphere, the Belt of Venus, and the Zodiacal light can only be seen with the naked eye. Meteors and meteor showers cover

On September 28th, 2015, in the early morning, Peter Horalek (seen in the photo) took this extraordinary photo while observing the Moon in Namibia's orange sand dunes. The Moon was passing through the Earth's shadow producing a spectacular total lunar eclipse which is seen at the center, close to the horizon. The dark oval in the sky is the Moon's shadow cast on interplanetary dust in the downwind tail of the Earth. Peter said his naked eye view of the heavens in this outstanding desert location was one of the top experiences of his life.
Credit: P Horalek

such a wide expanse of the sky that they too need the naked eye to see them. With your naked eye, using a solar filter, you can see large solar storms far larger than the Earth as black spots on the Sun moving across its face day after day when the Sun is very active, Lunar eclipses are best observed with the naked eye because you see a bronze Full Moon sitting amongst the stars. Similarly, a total solar eclipse impacts on the whole environment for as far as you can see, so they are well experienced with the naked eye.

OBSERVING WITH BINOCULARS

The naked eye can see a few thousand stars, but a pair of binoculars in a dark sky will reveal hundreds of thousands of stars. When looking along the Milky Way in binoculars, you will see an absolute myriad of faint stars.

Common 6x35 or 7x50 binoculars will reveal Jupiter's moons changing their positions hour by hour, and large 20x80 binoculars will start to show a hint of Saturn's rings. During the day, large sunspots can be seen easily in some detail in binoculars when using eyepiece projection, or by direct viewing using a solar filter. (See Volume 2 Chapter 3, page 66.) **Binoculars also give good wide-angle views of large features on the Moon such as its lava seas, its largest craters, and the ejecta rays** from large recent impacts.

These images simulate what the Moon and the Sun appear like when seen in small binoculars.

Because telescopes have much higher magnifications than binoculars, their field of view is correspondingly much smaller, so binoculars are better for observing large objects than telescopes are. Binoculars are best for **comets that have tails** that are too long to be seen in a telescopic field of view. They are perfect for seeing surprising detail in lingering **bright meteor trails**. During **total lunar eclipses**, the Moon looks more impressive in binoculars than in a telescope. And during a **total eclipse of the Sun** binoculars show the Sun's corona well, and also its prominences, but not in the detail that a telescope does. Large naked eye **star clusters like the Hyades and the Pleiades**, as well as galaxies like the **Large and Small Magellanic Clouds**

Mounted binoculars reveal far more detail than when they are hand-held. This is because out the slightest body movement blurs the image.

and the **Andromeda Galaxy** can be seen well in the wide fields that binoculars offer - much more so than in a telescope that only shows a part of these galaxies. Details in the **dust lanes of our Milky Way** can be seen very well in binoculars. In a dark sky, binoculars will reveal **the brightest stars in the closest star clusters. Large globular star clusters, nebulas, and galaxies appear** as obvious hazy spots in binoculars. After looking at the best deep sky objects a few times in binoculars, you will soon remember where they are relative to the naked eye stars in the constellation in which they reside.

You will be able to find over 200 deep sky objects of all kinds in binoculars, but most will be small and faint. If you use a computerized **Skyscout** on your binoculars, you will find them quickly, and you will be able to identify each one you find. (See Chapter 5, page 74 for details.)

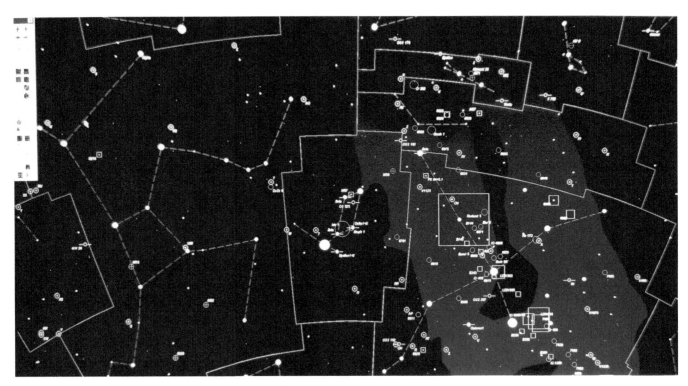

A section of the TUBA digital star atlas for binocular observing

A pair of binoculars are good for observing lunar occultations when the planets and bright stars go behind the Moon. Here we see Jupiter partially occulted by the Moon's dark limb. Two of its moons remain visible. Credit: Emil Ivanov

A GOOD BINOCULAR STAR ATLAS

Veteran observer **Philip Harrington**, produced the superb '**Touring The Universe Through Binoculars Atlas**' (TUBA). It's a digital star atlas for finding your way around the sky with binoculars. The concept was conceived by Dean Williams. It is available online for free! It is a very practical atlas for learning the sky using an iPad. This atlas has numerous very useful features.

TUBA shows stars to magnitude 11 making this atlas very suitable for use with large binoculars. The program allows you to reduce or expand the size of the field of view to match your binoculars field of view. You can have a field of view as wide as 70°, or you can zoom into just 1°. Zooming in to reduce the atlas' field size allows you to see fainter stars. The program is very versatile, and it's very practical to use. You can have white stars on a black background, which is best for observing. But, if you prefer, you can reverse this to have black stars on a white background for planning your observing session at your work desk. You can hide whatever features you don't need such as coordinates, constellation borders, the Milky Way, the location of the Moon, the planets and the Sun, or names. If you prefer a printed chart when observing, it allows you to print off any selection you make. Surprisingly, it is only for the northern hemisphere so the table of the top 20 deep sky objects does not include many superior objects in the southern sky.

BINOCULAR PARTS

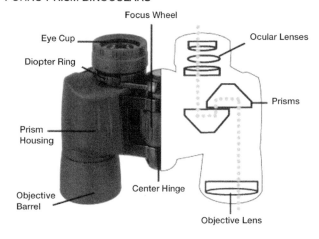

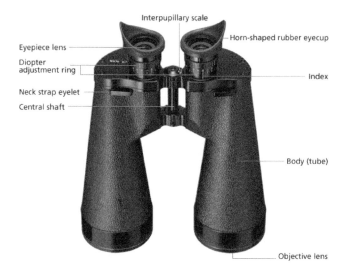

HOW BINOCULARS WORK

Binoculars are two small telescopes; one for each eye. They have two separate optical systems with each one consisting of an objective lens and an eyepiece. To shorten them, the optical path is folded back and forth using prisms.

With an optical system for both eyes, **binoculars provide a 3D image for terrestrial viewing**, whereas a telescope can only provide a 2D image. The more the main objective lenses are separated, the greater the 3D effect will be. Because the objective lenses of binoculars are further apart than our eyes, the three-dimensional view in binoculars is evident for objects up to a kilometer (~ a mile) away. In large binoculars, the 3D effect is noticeable up to around 10 km. The Moon and the stars are too far away to show any 3D effect.

Compared to monocular vision that uses only one optical system and one eye, binocular vision delivers a 40% increase in both detail and contrast because our brain is able to combine both images. Binoculars provide almost double the amount of information relative to the monocular view in a telescope. This makes the image clearer. Binocular vision is also more relaxing for the eyes - provided that the optics are properly aligned and focused so there is no eye strain.

A view through 200 mm (8y) binocular telescope is superior to that of a single telescope of the same size. Credit: Gregg Thompson

Testing the Advantages of 3D Binocular Vision

To test how powerful binocular vision is, use your naked eye to look into the distance where there is a range of objects (trees, buildings, telegraph poles, hills etc.) at varying distances from 100 meters to a few kilometers or so (100 yards to 300 yards or so). Ⓒ To test the effectiveness of your 3D vision, cover one eye with your hand. Make sure it does not press on your eye. Give your uncovered eye about 10 seconds to adjust to monocular vision and then try to determine how easy it is to work out which objects are in front of others. If you didn't know beforehand, you would have considerable trouble determining this as there is no perception of depth with one eye. Now, take your hand away s and it will become more obvious. Ⓒ Try walking around with a patch over one eye for a while. You will have trouble judging distances to things close at hand and those in the distance. When you take the patch off, you will realize how much better it is to have binocular vision.

The Finger-Touching Depth Perception Test

To demonstrate further the value of binocular vision, have a friend stand in front of you with th eir hand extended towards you with one finger held upright so that it is about half your fully extended arm's length away from you (Figure 1.). With both eyes open, hold your hand high above theirs and then quickly move your forefinger down so that it goes directly on top of their finger (Figure 2.). It's easy, isn't it? Now cover one eye and have them move their finger to a new position (Figure 3.). Now try to touch their fingertip using only one eye. With monocular vision, you will be unable to judge the distance to connect with their finger. This produces a lot of laughs. Let others try this and then explain why it is so difficult. We take our binocular vision too much for granted because, without it, we would have constant trouble negotiating our environment safely.

FOCUSING BINOCULARS FOR PERFECT VISION

To use binoculars, it's important to know how to focus them to get a sharp view in both eyes and to ensure you do not have double vision. Unfortunately, most people are never taught how to adjust binoculars. This is why so many people never use them much after they buy them. They typically stay hidden away in a cupboard. If binoculars are not adjusted correctly, they can strain your eyes and give you an uncomfortable feeling. Eliminate these problems by learning how to focus them correctly.

To focus binoculars, do the following:

1. Firstly, **move the eyepieces either closer together or further apart** until they sit comfortably in your eye sockets, so that you have a relaxed view. This will stop the risk of seeing a dark center or a dark edge. You should see only one round image - not a pair of overlapping views as shown in movies. If you do see double images, your binoculars may have been dropped and this has caused them to be knocked out of alignment. This is usually not the case because good binoculars are sturdy, so they can take quite a bit of knocking around. It may be that the objective lens on one side is not screwed in tight or that it is cross-threaded. If so, unscrew it and screw it back in so that it sits flat against the

body of the binocular. If the internal prisms have been knocked out of alignment, they will need to be adjusted by an optician at your local telescope or camera shop.

2. **Each lens system has to be focused separately** because many people have a different focus for each eye. To focus the left lens, place the lens cap over the right objective lens (or simply close your right eye) so you can only see through the left side while making sure that your *left eye is relaxed*. While doing so, focus the *left* eyepiece by turning the main focusing knob on the bridge of the binoculars. Turn it back and forth slowly until you get a perfectly sharp image. Focus on an object in the far distance.

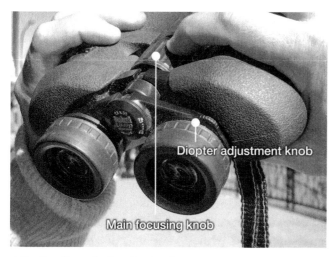

Adjusting the main focusing knob

3. Now, cover the left lens and look through the right side. Bring the right eyepiece to a perfect focus by turning the right **diopter knob/ring** on the eyepiece itself. (Do not touch the focusing knob on the bridge of your binoculars.) Some brands have a small lever to move on the right eyepiece. Obtain a sharp focus with your *right* eye ensuring it is relaxed. (On most brands, only the *right* eyepiece has its own separate eyepiece adjustment, but some brands have an adjustment on both eyepieces.)

4. Now, remove the right lens cap cover and look through your binoculars with both eyes. They should be perfectly focused. If not, repeat step 3 and make minor adjustments without replacing the lens caps this time - just ignore the side you are not adjusting. You should now have sharply focused images in both eyepieces and they should appear as a single image. When binoculars are set up correctly, your eyes easily superimpose both images to appear as one clear 3D image.

Adjusting the right hand eyepiece

If you don't bump your right eye focusing knob, from here on you will only need to use the main focusing knob on the bridge to change the focus from near objects to those in the far distance. If other people use your binoculars, they are likely to need to change your focus settings so they too may need to follow the procedure above. This is simple and quick to do once you have done it a couple of times. If they change your focus, simply readjust it for your own eyes after they have stopped using them. This only takes about 10 seconds to do. There are a number of videos on the Internet that explain this procedure.

Cheap binoculars often have weak eyepiece support arms for the eyepieces, so they can wobble. This allows them to be easily knocked out of focus. If yours are like this, it's best to press both eyepieces forward to level them before looking through them, and then adjust the focus.

Use the **neck strap** so you don't risk having your binoculars accidentally knocked out of your hand. It only takes a couple of seconds to take this precaution.

When they are not in use, make a habit of keeping the **dust caps** on the eyepieces. Place your binoculars on a convenient shelf with the objective lens down so they don't catch dust or sea air. Dust and moisture will dull the image causing a loss of clarity, and it will make the image foggy thereby considerably reducing contrast. This makes shadows become washed out, so detail is lost. Greasy lenses will cause flares from bright lights and stars. **The objective lens and the eyepieces must be clean to ensure a sharp image.** If the lenses become smudged with sea air or dust, clean them with *very mild* detergent on a damp, clean, soft cloth. Do not use tissues as they scratch lenses and they often leave smudges. A cleaning cloth is usually provided with your binoculars, so use it. These cloths will need to be washed out after several uses.

If you bought your binoculars to use regularly for terrestrial viewing as well as astronomy, then *keep them where you can grab them quickly* when there is something interesting to look at. This way, you will be likely to use them regularly and you will get your money's worth from them. If you pack them away in their box and you put the box in a cupboard where it is out of sight, then you are unlikely to ever use them. Remember that they show far more detail than the naked eye can see so you will be wise to have them handy so you can use them. If your binoculars have a magnification of 7 times, then you will see 7 times as much as you can with your naked eye, providing you keep them steady.

THE FIELD OF VIEW

The field of view of a pair of binoculars is typically printed on the binoculars' wheel plate or their side. It is usually given as a linear measurement stated as the number of meters wide that you can see at 1,000 meters distance, or it is given as the number of feet wide at 1,000 yards. It is sometimes given in angular degrees, which is better for astronomers.

BINOCULAR SIZES

Binocular specifications are denoted by pairs of numbers such as 8 x 35, 7 x 50, and 20 x 80. The first figure refers to the magnification and the second number refers to the aperture of the front (objective) lens measured in millimeters. The larger this second number is, the more light the lens will gather, so the more stars and nebulas you will see.

Very large binoculars are like two spotting telescopes that can magnify 20-40 times. Remember that the larger the magnification, the larger the binoculars are, and therefore the heavier they are, and the more expensive they will be. Because large binoculars are too heavy and bulky to be hand-held, they require an expensive mount to support them. Once binoculars magnify over 7 times, they are difficult to hand-hold because the slightest, unintended muscle movement is magnified over 7 times. This causes a significant loss of detail. To utilize the ability of large binoculars to show much detail, they must be mounted.

The weight and bulkiness of binoculars should be considered if you want to use them when traveling, hiking, or at sports events and concerts. Small binoculars, or a monocular, are excellent for concerts where you are not close to the stage. You can see a singer or others up close and see their expressions as if you are in the front row. For such uses, small, lightweight binoculars are the most practical. Small binoculars are typically used where there is plenty of light, so there is no need to have large lenses. Large objective lenses are only helpful for stargazing. But small binoculars with 35 mm objective lens still work fairly well for astronomy.

When hiking, rather than carry bulky binoculars, I carry the very practical, small, lightweight 8 x 30 Pentax monocular telescope that has a 6° field. It also acts as a 32x magnification microscope for looking at flowers, seeds, mosses, fungi, and insects etc. When these things are viewed in sunlight, you see truly remarkable views of the micro world. You simply screw a macro lens onto the front of the objective lens.

The macro lens can also be used on its own as a powerful 4 times magnifying glass. (It is shown on page 131.) The Pentax version is no longer available but Barska sell an inexpensive 8 x 25 monocular telescope/microscope for a very reasonable US $35.

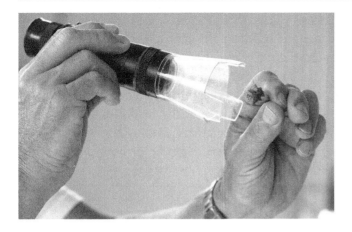

This is a small 8 x 30 spotting pocket telescope. It has a screw-on 4x magnifying lens. When used with the spacer (clear piece), it becomes a superb microscope with a magnification of 32x for looking at things in the macro world. When small flowers and insects are placed in sunlight, they reveal spell-binding textures, details, and fluoro colors. This combination of a lightweight, small telescope and microscope is a superb accompaniment when you are in nature.

Left: **From largest to smallest, the author's much used old Kassel 20 x 80s for deep sly observing give a magnification of 20x. The Bushnell 12 x 50s provide a magnification of 12x for general viewing. The inexpensive pair of lightweight Celestron 7 x 35s have a magnification of 7x, which is good for terrestrial viewing and for traveling where a lightweight, small pair is practical. Because they are inexpensive, if they go missing or get damaged, it's not a big loss.**

Center: **Giant 150mm (6") naval binoculars from an old warship provide a magnification of 30x and high definition.**

Right: **Nikon 20 x 120 are high quality, 20x magnification, sturdily mounted binoculars used on a cruise ship.**

LARGE BINOCULARS

Large binoculars deliver superior views of many objects due to their large objective lenses and their high magnification. They are essentially a small telescope for each eye. This pair is mounted on a good quality video camera tripod mount. This type of mount is good for terrestrial viewing, however, it is very uncomfortable for viewing objects above 30° above the horizon because this causes back and neck strain.

The larger the objective lenses of a pair of binoculars are, the more light they will gather and the more they will magnify to show fainter objects and more detail. But the larger the lenses are, the heavier they are, and the higher the magnification, so they must be mounted to hold them steady.

BINOCULAR MOUNTS FOR COMFORTABLE OBSERVING

Binoculars with a magnification over 7 times have to be mounted to see fine detail, because if they are hand-held, tremors in the body cause fine detail to be lost due to the image constantly moving around. To help keep even small binoculars still when they are hand-held, it helps to brace your arms by putting your elbows tight against your chest, or better still, lean back against something solid like a wall. Resting your elbows, or the binoculars themselves on something solid like a rail or a low wall will steady them. It's best however, to fix them to a solid, adjustable-height camera tripod. You can buy a small, inexpensive binocular holder that screws into the bridge of the binoculars and the other end fits onto a camera tripod. These items can be bought from most camera and telescope stores.

To overcome neck strain when using binoculars for astronomy, lie on a sun lounge with an adjustable back. Gary Seronik adapted a camera tripod to create this relatively simple and inexpensive mount for small binoculars. Details of how to build one were published by Sky & Telescope magazine. This excellent magazine is available online to subscribers. Credit: G Seronik

When looking at objects high in the sky, it's best to lie down because it only takes less than a minute before you start to get neck strain and/or a headache by having your neck bent backwards. As well, it becomes uncomfortable to have your arms to be holding up binoculars for more than a minute. A solution is to have your binoculars look downward into a good-quality mirror that allows you to adjust the angle of the mirror to look at different parts of the sky. This provides a comfortable viewing angle. Astronomy magazines and the Internet have advertisements for such mounts that have a built-in swivel mirror.

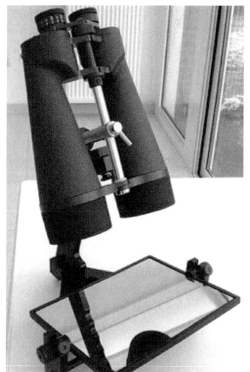 

Left: **For comfortable viewing with large binoculars, use this commercial mounting so you can sit in a chair and place the mirror mount on a low coffee table so you can to look down into the binoculars comfortably while peering into a reflection of the sky above. You can move around the sky by tilting the mirror and rotating the stand.**

Right: **Counterbalanced mounts for large binoculars provide great stability necessary for viewing large, bright, deep sky objects and comets, however one can quickly develop neck strain with this type of mount when viewing objects high in the sky.** Credit: Dennis di Cicco

 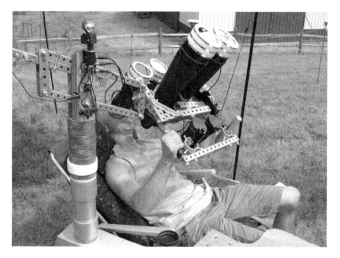

Left: **This style of binocular mounting counterbalances the weight of the binoculars and it keeps them steady when the observer is reclining in a sun lounge. The disadvantage with this setup is that you have to keep moving the sun lounge to move around the sky.**

Right: **Roger Greenwood seen here, built this homemade, multi-binocular mount complete with a motor drive and a handy bench top. This mount also requires the chair to be moved each time one wants to move around the sky. Note that the mount carries 3 different pairs of binoculars! In this photo, he is observing the Sun using aluminized Mylar solar filters.**

A ROTATING BINOCULAR CHAIR

I gave considerable thought as to how to design a practical binocular mount that would overcome all the problems that other binocular mounts have. I came up with the **Thompson Rotating Binocular Chair**. It is a superb solution for supporting binoculars of all sizes, particularly large ones. The basic version shown here provides the most practical solution for making an inexpensive, vibration-free binocular mount that can look in any direction, and at any altitude without the observer needing to move from his chair. And because it allows the observer to be very comfortable no matter where he looks in the sky, there is no risk of getting back or neck strain. The chair can recline right back when one wants to look directly overhead, and it sits upright for terrestrial viewing.

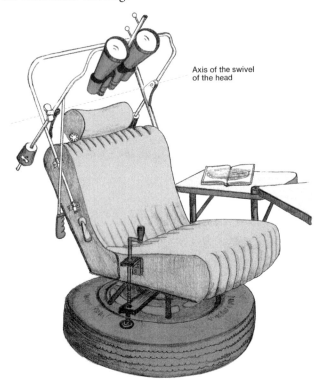

This is my original rough drawing of my rotating chair design, which is the most practical and comfortable of all binocular support systems.

This rotating chair design is the most practical and comfortable binocular support system.

To make a Rotating Binocular Observing Chair inexpensively, follow these steps:

1. Purchase a second-hand car seat – one that allows the seat to lay right back by adjusting a lever on the side, or better still car seat. Do not use a screw type adjustment because they are too slow and far too difficult to change the backrest angle. The seat must have a headrest.

2. Using four steel rods, weld the chair base onto a second-hand car wheel and bearing assembly. Inflate the tire for stability. Car seats and wheel assemblies are available at car wrecking yards. The wheel's bearing permits the seat to turn easily in the **horizontal (azimuth) axis**. When seated, you can control the degree of movement in the horizontal very accurately by simply moving your feet on the floor or the tire. The car wheel's bearing provides smooth rotation that allows fine movements in the horizontal axis.

3. On the sides of the seat, fix a length of metal tubing bent into shape as per the drawing so that the top arches over about 250mm (10") above where the top of the observer's headrests. You can use thin aluminum tubing as it is easy to bend. This frame will support the **altitude arm** that supports the binoculars. This assembly allows the binoculars to move up and down with ease. Make sure the altitude arm protrudes forward far enough so that when you get in and out of the chair, it can be lifted up and back so that it is well out of the way of your head.

4. Weld a small **support plate** to the vertical tubing on either side to mount the altitude arm. Drill a hole through the plate at the point where your head swivels back and forth on your neck. (See below.) A bolt goes through this hole and also through a hole in the altitude arm to act as the axis for the altitude arm. This acts as a bearing to allow the altitude arm to move up and down. There needs to be just enough friction to hold the binoculars at the altitude that you want to observe without the binoculars moving up and down too freely. If you want to change the angle to go well above or below this, then adjust the back of the seat to suit. There is no need for a bearing at this **swivel point**. It's important to locate the support bracket swivel point *on the line that passes through the pivot point of the observer's neck*. This is at the middle of the nape of the neck. This keeps the binoculars at the same distance from the observer's eyes when the observer's head changes position for a range of altitudes that are consistent with the angle of the seat's backrest. You do not need to adjust the angle of the back seat until you need a significant

change in altitude. The axis of the swivel point of an observer's head must be at the right height, and also the right distance from the back of the seat when the observer has his head resting on the headrest. This need to be measured when an observer is sitting in the chair. If short people, or children use the seat, it will work best if they use a pillow to sit on to place their neck at the swivel point of an average person. There is considerable latitude in this, so it will not be a problem for tall people.

5. Weld a straight piece of tubing about 200 mm long to the underside at the center of the front of the altitude arm so that it protrudes forward. Select a piece of metal rod about 400 mm long that snugly slides inside this tube but still allows it to slide with ease. The rod's ability to slide allows the binoculars to move forward when the observer wants to get in or out of the chair. It also allows for some slight in and out adjustment of the binoculars to accommodate for small changes in the distance from the observer's eyes when the observer changes the altitude of the binoculars without changing the angle of the chair's back. To the lower end of the **sliding rod**, attach **a clamp** that will tighten onto the focusing bridge of the binoculars.

6. So that you do not have to hold your arm up to move the binoculars, attach **a swinging lever** to the side of the altitude arm to move the binoculars in the vertical axis.

7. Attach sliding, adjustable **counterweights** to the rear end of the altitude arm frame to balance the weight of the binoculars. These can be moved to counterbalance any pair of binoculars.

8. The binocular chair's tire can sit on any flat surface. If you need to move the chair around or place it undercover when it is not in use, fit dolly wheels to the underside of the wheelbase. Alternatively, have a vinyl cover made to cover the seat so you can leave it outdoors. The binoculars are unclipped and taken inside when the chair is not in use.

9. A couple of adjustable viewfinder pins can be welded to the top of the sliding tube holder bar to act as sights when you sit behind the chair and hold onto the counterweights to maneuver the binoculars from one object to another to give passengers a tour of the sky. I have often shown visitors bright deep sky objects and the galaxy's dust lanes this way. They say it is like going on a voyage in a spaceship!

When you are comfortably seated for observing in the chair, pull the altitude arm down and slide the binoculars towards you until the eyepieces are comfortable against your eyes while your head is supported by the headrest.

To **demonstrate the conservation of energy** when a star contracts, it is fun to place people in the chair sitting upright with their arms and legs extended and then spin them slowly. Then ask them to slowly bring their arms and legs in close to their body. As they do this, they speed up considerably. When they extend their arms and legs, they slow down. They can do this a number of times on one spin. Because this demonstration is so dramatic, it fascinates people, and it's a lot of fun.

In a commercial stargazing venue, the wheel assembly could sit on a platform. There could be a number of binocular chairs.

The azimuth axis on the wheelbase and declination axis on the arm could be motorized to move around the sky from object to object just as a GOTO telescope does. This would provide an interstellar 'joyride' to those in the chair without them having to touch the binoculars. The chairs could be programmed to have several 'rides' – one or two for each season. During each ride, the binoculars would go a number of objects. If there is cloud in one part of the sky hiding some objects, then the controller could delete those objects and add in others that are visible. The binoculars would automatically move in unison from one selected object to another. The chairs would be car seats with motorized recliners. They would be programmed to move to each object's altitude above the horizon at that time.

Observers could wear headphones that have a narration about each object together with music to take 'space voyagers' on their exciting unique interstellar joyride. Alternatively, open air speakers could be used. Each object would have a short statement when it was acquired, or, the supervisor would narrate the show. To make it more exciting, at the cardinal points surrounding the binocular chairs, images of each object as seen in telescopes could be projected onto 4 outdoor screens sitting above the head height of onlookers. The images would appear after the observers viewed each object in their binoculars. After this, they would automatically move onto the next object.

The rotating binocular chair is a very functional means of supporting a pair of large binoculars as well as your head. It allows comfortable viewing of the whole sky.

TEST THE QUALITY OF BINOCULARS BEFORE BUYING

There are some binoculars that are virtually useless. This is often the case for cheap plastic ones, and those for children, which are a total waste of money. Some large, cheap zoom binoculars can look impressive but the moment you look through them, you will notice some of the following problems:

1. The image is a little blurry even at the center of vision, and the clarity worsens towards the edges. This is caused by poor quality lenses.
2. The image is somewhat dark. This is typically due to the objective lens being too small for the magnification used.
3. There is a very small field of view that gives the impression of looking down a tube. This is due to the use of poor quality optics that have blurry edges, so they have been blocked off.
4. They are hard to focus and to keep in focus. This is caused by the eyepiece assembly arrangement being flimsy allowing it to wobble around.
5. Objects have strong color fringes. This is due to cheap, poor quality lenses.

If you notice any of these problems, don't buy the binoculars.

When buying binoculars, the price should not be the deciding factor. Just because some binoculars are expensive does not mean they are always good for your needs. And conversely, some relatively inexpensive binoculars can often perform quite satisfactorily. Don't be seduced by high magnification binoculars or zoom lenses, especially if they are at the low end of the price range. These are usually of very poor quality. A good pair of binoculars will give you a sharp view that is clear at least halfway to the edge of the field of view.

Before buying binoculars, *look through them to check out their quality.* **This is the only real test.**

Some binoculars boast anti-glare coatings but this feature has very limited value. Some are waterproof but this too is only of value if you think you could drop them over the side of a boat or be in the rain when hiking. You are best to start out with a modest, inexpensive pair with a magnification of 6 to 8 and a lens diameter of 35 mm to 70 mm (1¼" to 2¾"). Once you have experience in using binoculars, then look at what you might gain by buying a very good quality pair or a large, mounted pair.

In the next chapter, we will look at tips for using and buying telescopes.

Chapter 9
USING AND BUYING TELESCOPES

The marvelous invention of the telescope has allowed us to see amazing detail in the universe. A telescope not only allows us to greatly magnify objects in space but it also brightens our view of very faint objects.

THE WONDER OF THE UNIVERSE REVEALED

When Galileo used his first telescope for observing objects in the sky, he opened a window into the heavens that allowed us to see so much more than was possible with our naked eye. Thanks to amateur telescopes, the planets are no longer stars: we see them as other worlds. Telescopic cameras on space probes have given us the most incredible views of features on other worlds that were not previously imagined. And the Hubble Space Telescope has revealed the most beautiful, unexpected, truly stunning images of the universe. It has allowed us to see almost back to the beginning of time! Much of the majesty of the universe that is presented in *STARGAZING* is only available to us due to the wonderful invention of the telescope. The value of this instrument cannot be overstated.

The Very Large Telescopes at the European Southern Observatory in Chile is so large and advanced that is like something out of a science fiction movie.

Today's huge telescopes used by professional astronomers are so expensive to build, operate, and maintain, that astronomers wanting to use them for their research programs have to put forward compelling scientific research proposals to justify their use of these monster telescopes. By contrast, amateurs using their own telescopes can observe whenever they want to, and whatever they like. They can observe the sky purely for the enjoyment of looking at many fascinating objects, or they can do serious research that can assist professional astronomers. Others love using their telescopes to take breathtaking astrophotography.

Einstein thought that Galileo was the father of all the sciences because he made us understand that the contemplation of nature alone was useless: only observation of it would deliver an accurate view of how nature works. With his telescope, Galileo created the means by which we could observe the universe to discover what it contained and how it worked. With great insight and intellect, Galileo stated that if there was a creator that existed outside of the universe, then such a creator would be unobservable and therefore outside of the field of science. A creator could only be conjecture with no means of proving its existence. As true and obvious as his statement was, this did not endear him to the Church.

In 1620, Galileo was the first person to use a telescope to peer into the universe.

Galileo found out that Dutch glass makers had invented lenses that could make distant objects appear closer. In 1606, he purchased some of the Dutch lenses and made himself the first telescope. He thought it would be useful for watching which ships at sea were coming in to dock for commerce. And it was also useful for military applications, but in the end, it became the most useful for astronomy and photography. The Senate of Venice was so impressed with Galileo's telescope that he received an increase in his salary and tenure for life at the Padua University as chair of mathematics. Using his telescope for astronomy, he discovered mountains and craters on the Moon, the four large moons of Jupiter, sunspots on the Sun, and the Sun's rotation. He also observed that Venus had phases and that the size of its disc changed. From this he cleverly deduced that the only way this could occur was due to Venus orbiting the Sun inside the Earth's orbit. The genius of Galileo had the profound effect of changing our beliefs and perspectives on so many things.

TYPES OF TELESCOPES

If you are thinking of buying a telescope, it's best to first learn about the basic advantages and disadvantages of the main types of telescopes before you start spending your money on one. There are two main types:

1. *Refracting* telescopes, which refract light passing through the main lens to a focus. Galileo's small telescope used this design.
2. *Reflecting* telescopes that use mirrors. These are based on Newton's telescope design. (See image on page 141.)

Let's now look at both these types of telescope designs in a little detail.

REFRACTING TELESCOPES

The original type of telescope that Galileo invented is called a **refractor**. Today they are more practical for terrestrial viewing than for astronomy. When looking at the stars with a refractor, they require a diagonal prism to bend starlight at 90° to facilitate comfortable viewing.

Once the objective lens of a refractor exceeds a diameter of 60 mm (2.5"), they become very expensive, particularly if they are of a good quality. Objective lenses that are comprised of only one lens suffer from **chromatic aberration** – a problem where the light of different colors comes to a slightly different focus. This causes the image to have blue colorization on one side and red on the other thereby blurring the image somewhat, particularly in the outer parts of the field. This occurs because the edges of a lens act like a prism breaking white light into its component colors. This can be overcome to a considerable degree by using expensive, **compound lenses** consisting of two or more lenses made of different types of glass with different refractive indexes.

A good quality refractor with an aperture of 60 mm (2.4") to 90 mm (3.6") with a sturdy mount and a good viewfinder, as well as slow motion controls, and a couple of fairly good eyepieces will cost between US$250 and US$900. The price will depend on the quality of the objective lens and the eyepieces, as well as how sophisticated and robust the mount is.

NEWTONIAN REFLECTING TELESCOPES

The first reflecting telescope was built by Sir Isaac Newton in 1668 however, it was actually Galileo that first suggested making a telescope using mirrors. Newton ground an alloy of tin and copper into the shape of part of a sphere for his main mirror. This would bring all the light gathered by the mirror to one focus. His first telescope had an aperture of 30 mm (1.3") and a focal ratio of f5 (150 mm).

A relatively good quality reflecting telescope with an aperture of between 150 mm (6") up to a 250 mm (10") with an Equatorial or Dobsonian mount without a motor drive will cost between US$300 to $800. Cost depends on the quality of the mirror and the eyepieces, and the type of mounting the telescope has. A large 310 mm (12.5") Dobsonian reflector can be purchased from around US$1,000. A very large equatorial mounted 450 mm (16") reflector costs US$2,000 to US$3,000. These prices represent very good value for money. Adding a computer-controlled motor drive to a telescope increases the cost by a few hundred dollars. It provides a very valuable benefit that permits easy and fast, computer-aided acquisition of objects and automatic tracking of them as they move across the sky due to the Earth's rotation. The relatively small cost of having your telescope computer-controlled is worth every cent because it gives you more observing time, and it makes observing much more relaxing and pleasurable compared to manually finding objects and tracking them.

SCHMIDT-CASSEGRAIN REFLECTING TELESCOPES

Schmidt-Cassegrain or Catadioptric telescopes are reflecting telescopes that employ an improved optical design to achieve high-quality images and short tube lengths. Because of this, they are more expensive than Newtonian reflectors. The design is compact and relatively lightweight for their size, making them easy to transport and carry. Their compact, lightweight nature makes them easy to set up and maneuver. This makes them more likely to be used than a heavy, bulky telescope.

Because Schmidt Cassegrains have the eyepiece located at the bottom of the telescope behind the main mirror, you can often sit in an adjustable height chair to observe through the eyepiece, whereas with a Newtonian design, you typically stand to look through the eyepiece. When looking at objects high in the sky with a large aperture Newtonian telescope, the eyepiece is so high as to require a ladder to reach it.

These images show the evolution of the reflecting telescope.

Left: *Isaac Newton's small 33 mm reflecting telescope,*

Center: *NASA's 2.4 m Hubble Space Telescope*

Right: *Tomorrow's giant 30 m telescope presently under construction at Hawaii by the TMT Telescope Corp.*

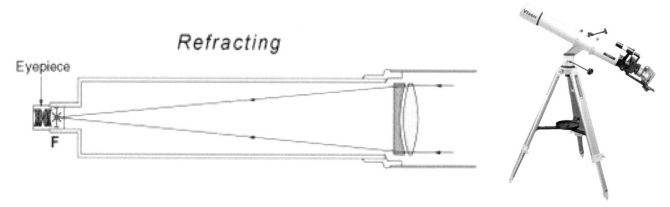

A refracting telescope gathers light using a convex objective lens at the front of the telescope tube. This lens concentrates light to a focus at the eyepiece. Here, interchangeable eyepieces are used to provide a range of magnifications.

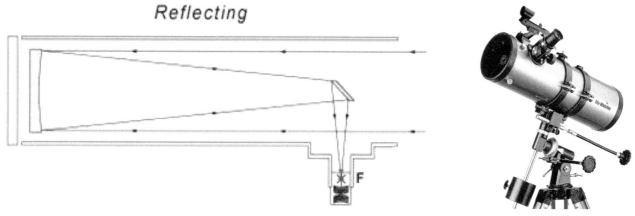

A reflecting telescope uses an aluminized concave primary mirror at the base of the tube to collect light which it reflects back up the tube to a small, flat secondary mirror at the open end of the tube. This mirror reflects the light out of the tube to an eyepiece holder at the focus. Eyepieces are placed here to magnify the primary mirror's low magnification image.

Schmidt-Cassegrain Telescope

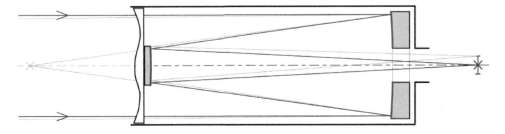

The Schmidt Cassegrain telescope is a conveniently-shorter version of a Newtonian Reflector. It uses a concave lens at the front of the tube to correct for its short focal length. A secondary mirror is placed at the center of the objective lens to reflect light back through a hole in the center of the main mirror to the focus, which is behind the main mirror. Eyepieces are placed here to magnify the primary mirror's image.

Schmidt Cassegrain brands have good quality motor drives that have slow motion controls and GOTO computerized object-finding and tracking technology. These types of telescopes are a delight to use. A 200 mm (8") aperture will cost around a very reasonable US$3,000 while a very large 360 mm (14") aperture will be around US$10,000. These costs are exceptionally good value for such fine telescopes. Because they are a closed design, the internal optics stay clean longer.

BUYING A DREAM OR A NIGHTMARE?

There are many brands of telescopes most of which are well made and of good value, but there are a few very cheap brands that are designed to suck in unsuspecting novices. Buying one of these poorly made telescopes for someone is the best way to unintentionally destroy their interest in astronomy while at the same time wasting your money! Many well-intentioned people purchase cheap telescopes as gifts for their excited children or a family member without having researched what is the best telescope to buy. Cheap telescopes can look good to someone who knows nothing about them, but they are close to useless when you try to use them. The recipient often finds that on their first attempt to use such a telescope, it is a frustrating experience. After one or two disappointing attempts, the telescope typically ends up under the bed and it stays there.

Cheap telescopes have a mount that is so unstable that the slightest nudge or breeze makes it vibrate badly thereby blurring the image. The viewfinder is so inadequate that it is very difficult to line up objects that you want to look at. The eyepieces are of such poor quality that their small fields of view look like you are looking down a drainpipe and much of the field is blurry. Because many parts of cheap telescopes are made of fragile plastics or weak metal alloys, they break easily.

If you end up with one of these, you would be advised to return it and get a refund so you can purchase a good telescope. So you don't risk buying an unsuitable telescope, it is worth spending a little time researching what the advantages and disadvantages of different types of telescopes are, as provided here.

Some large camera shops have fairly cheap telescopes for sale and their staff are typically not well-informed about them, as they have probably never used one. It's best to go to a specialist telescope shop that has many telescopes on display and experienced staff who can advise you on what is best for your needs and budget.

To be assured of quality, it is important to buy a well-known, reputable brand. You can often get a good quality, second-hand telescope on eBay or similar sites at very low cost, but be sure to test it before purchasing it. If you don't know enough about telescopes, don't be too shy to ask an experienced amateur astronomer in a local astronomy club for assistance. Most will be happy to share their knowledge with you. Nearly all clubs are listed in the phone book or on the Internet. There are a number of sites on the Internet that have short videos that explain how different telescopes work and what you can expect to see with them.

Unless you are technically minded enough to build your own telescope, it is usually best to start with a simple, easy-to-operate telescope such as a 200 mm (8") or 250 mm (10") Dobsonian-mounted Newtonian reflector. These telescopes have inexpensive mounts, which are very stable. As you advance, you might want to upgrade to a larger telescope. You will be able to sell your existing one fairly easily, if need be.

Cheap 60 mm refractors are generally of poor quality, so they are to be avoided, but good brands are fine because they have good optics, solid mounts, good viewfinders, and they often have slow motion controls. For what you pay for a good brand of refractor you can buy a much larger aperture reflecting telescope that will enable you to see much more.

Low-cost telescopes are manually operated, so to find any faint object, you have to look through a small finder telescope and star-hop to the object's position as shown on a star chart. To do this, can take some time to learn how to do this. If it is a faint object and there are no bright stars nearby to hop from, this can be tedious, and it often requires some strenuous bending. For a relatively small additional cost (several hundred dollars), you can purchase a **computer controlled**, motorized telescope that automatically finds objects for you and automatically tracks them to compensate for the Earth's rotation. A motor drive will keep an object in the center of your field of view, thereby providing maximum clarity. Without a drive, objects drift across the field, so you have to keep nudging the telescope a little to keep the object centered.

Several decades ago, most amateur astronomers had to build their own telescopes because commercially available telescopes with respectable apertures were very expensive. Nowadays, telescopes with large apertures that are of high quality and , motor-driven commercial telescopes are very affordable, so they are an obvious choice for most amateurs, as well as for institutions such as schools and eco-resorts.

If you live in suburbia, it's a good idea to purchase a telescope that is easy to transport by car to dark observing sites, or to astronomy club field nights. Most telescopes can be easily disassembled for transportation and reassembled within 10 minutes.

Once you have gained experience with a medium-sized telescope, you may then have enough experience to justify buying, or even building, a larger instrument.

THE BASICS OF A TELESCOPE

The following points are things you should know about a telescope:

1. The aperture (diameter) of the main mirror of a reflector, or the objective lens of a refractor. This lets you know how much light it will gather and how much detail it will show if it is a well-made mirror. Mirrors today are machine made and computer controlled so most are of good quality.
2. The optical quality of the main mirror, or objective lens, so you know how much detail the image will show at high power under good seeing.
3. The focal length so you know how to calculate the magnification it produces when using different eyepieces.
4. The design so you know how long the tube will be.
5. The brand, if it is a commercially manufactured telescope.
6. The type and brand of the eyepieces so you know how good they will be and how wide their field is.
7. Whether it has a motor drive to track objects or a computerized 'GOTO' system which will automatically find objects for you.
8. The type of mount it has so you know if it will be stable and easy to use, or bulky and heavy. (See page 149.)

The definitions of each of these components, their specifications, and what each one means for the operation of your telescope are discussed below.

APERTURE

Aperture is the term that refers to the diameter of a refracting telescope's objective lens (the one at the front) or a reflecting telescope's main mirror (the one at the bottom). The objective lens or mirror is what gathers light. The larger the aperture, the more light the telescope will collect so the brighter the image will be. With a good quality mirror, the larger the aperture, the clearer the image will be, so the more it will stand high magnification.

In the *STARGAZING* volumes, when specific apertures are not given, telescope apertures are classified into the following ranges by size.

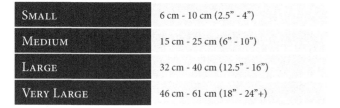

SMALL	6 cm - 10 cm (2.5" - 4")
MEDIUM	15 cm - 25 cm (6" - 10")
LARGE	32 cm - 40 cm (12.5" - 16")
VERY LARGE	46 cm - 61 cm (18" - 24"+)

OPTICAL QUALITY

Most good quality telescope manufacturers define the quality of their optics in terms of a fraction of the wavelength of light. For a telescope to be able to magnify well and produce a sharp image, the primary mirror needs to be figured so that the maximum surface variation is no greater than $\frac{1}{8}^{th}$ of the wavelength of light from one wave crest to the next. The very best quality mirrors have surface variations that do not exceed $\frac{1}{20}^{th}$ of a wavelength. This is a very small error indeed. Average quality mirrors typically have an accuracy of around $\frac{1}{10}^{th}$ to $\frac{1}{8}^{th}$ of a wavelength.

In the past, some amateur observers have had to return their main mirrors to the supplier for refiguring because the optics were not of the quality guaranteed in the marketing material. Recently, this has become much less likely to occur because modern Schmidt-Cassegrain telescope optics are almost always of a high quality since they are machine-made, so they meet their stated specifications. It is unusual nowadays to have to return a mirror for refiguring.

GOOD COLLIMATION

The term **collimation** - also known as **alignment** - refers to the process of ensuring that the optics of a telescope are correctly aligned along the focal plane. When a telescope is correctly collimated, the image is at its sharpest at the center of the **optical axis** (i.e. at the center of the eyepiece - providing the telescope is well collimated).

Outside this central area, distortion gradually increases. This distortion, or **aberration** as it is known, is referred to as **coma**. Very short focal length telescopes are most prone to having distorted images due to imperfect alignment of the optics. A short focal length telescope with a f/4.5 mirror has to be perfectly aligned to within 2 mm of the focal plane whereas a long focal length mirror of f/10 has an error margin up to 22 mm. If a telescope is not collimated well, it will display a distorted image even at the center of the field of view, especially at medium to high powers. Poor collimation produces flares to one side making stars look oval. This results in a loss of detail.

I will now cover the basic approaches to collimating different types of telescope. You can find good articles and videos on the Internet that demonstrate how to go about collimating a telescope.

COLLIMATING A NEWTONIAN TELESCOPE

A Newtonian telescope has three adjusting screws at the back of the main mirror cell. These allow you to tilt the main mirror to align it with the small, flat, **secondary mirror** (also known as a **diagonal mirror**). The secondary mirror should be positioned at the center of the telescope's tube so that you can see the whole of the main mirror reflecting in the secondary mirror when you look through the eyepiece holder (when the eyepiece is removed). There are three adjusting screws on the back of the secondary mirror holder to allow you to align the reflection of the main mirror so that it is in the center of the diagonal mirror when you look through the center of the eyepiece draw tube. The eyepiece holder used for focusing must be at a right angle to the telescope tube and it must be centered on the optical axis. If you do not have a motor drive, when aligning any telescope, be sure to move the image of the star back to the very center of the eyepiece before making the next adjustment.

A Schmidt Cassegrain telescope has three adjusting screws on the front of the secondary mirror holder. They seldom need adjusting but if they do, make small, $\frac{1}{8}^{th}$ turns clockwise or anti-clockwise to center the out-of-focus image of a star in the eyepiece. You need to align the mirror so that stars have perfectly concentric diffraction circles around them, as shown in the following diagram.

A **Star Test** is the best way to test the optical quality of your telescope and the telescope's **collimation**, but it does require some experience. This test will assist you in correctly aligning your optics to get rid of flares and blurred images. Internet sites will also show you how to interpret the shape and the brightness distribution of a star's light at the focus as well as inside and outside the focus.

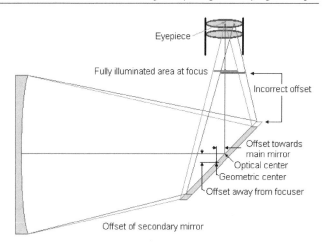

A correctly aligned optical path is shown in gray and misaligned one in red.

The poorly aligned red-light path is losing light around the outside of the secondary mirror and it comes to a focus that is not aligned with the center of the eyepiece. This will cause the image to be distorted.

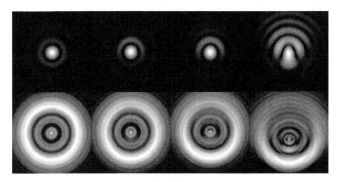

The star test shows whether your telescope is correctly collimated.

The image on the left in both rows shows good collimation, whereas those moving to the right show increasingly poorer alignment of the optics. This causes coma and blurred images. The top row shows a simulation of a star's image under perfect seeing at the best focus. The bottom row shows the image of a star just outside the focus.

inside focus ——— focused ——— outside focus
Overcorrection in the presence of turbulence

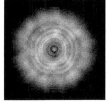 

Same overcorrection under good seeing

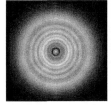  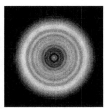

How a medium brightness star will appear when looking through a medium power eyepiece during poor seeing and good seeing.

Reflector telescopes use a large concave mirror at one end of the tube, and a small flat mirror at the other end. If the secondary mirror is not centered on the converging light beam from the main mirror, then this will cause poor collimation. There can also be a loss of light coming from the main mirror if the secondary mirror is placed too close to the main mirror, or it is off to one side. In this case, light from the main mirror goes around the edges of the secondary mirror and is lost. If the eyepiece holder is not aligned square to the optical path, this will cause poor collimation. The shorter the focal length, the more accurate the collimation has to be. There are **Cheshire eyepieces** and **laser collimators** to assist with collimation. There are videos on YouTube explaining how to use them, but these are not necessary if you follow the instructions given here.

Refractors use fixed lenses that are factory collimated so it is very rare for them to need re-collimation unless they have been badly knocked and noticeably damaged. Catadioptric (SCT) telescopes may occasionally require collimation of their secondary mirror after rough transport. Due to the design of Newtonian reflecting telescopes, they often require regular re-collimation, particularly after being transported over rough roads to a dark sky site, or from being regularly moved in and out of a house.

FOCAL RATIO & FOCAL LENGTH

You will hear the words 'focal ratio' and 'focal length' mentioned when talking about telescopes. The *focal length* of a telescope is the length in mm between a telescope's primary lens or mirror, and the point where the light rays come to a focus. Commercial telescopes display the focal length on the telescope tube.

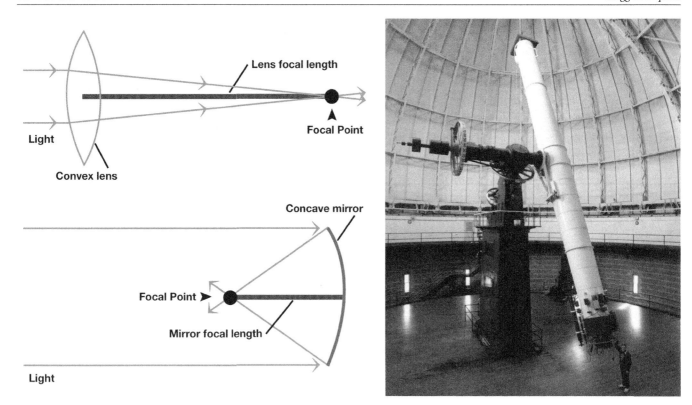

The focal length of a telescope is the distance required to bring the light rays to a focus.

Left: The diagram on the left shows the light path through the objective lens of a refracting telescope while the lower diagram shows the light path of a reflecting telescope. The light paths have been greatly shortened for the illustration.

Right: The 40" Yerkes telescope at the University of Chicago is the largest refractor in the world. Its focal length is 19 m (63'). Telescopes this large require very expensive domes to house them. The lenses of very large refractors are so heavy that they can slowly sag over many years under their own weight. This causes distortion of the image, so the telescope has to be left in different positions when not in use in order to try to minimize sagging.

You can calculate a telescope's focal length by multiplying the diameter of the main lens or mirror by the telescope's focal (f) ratio. For example, a 200 mm (8") telescope with a f-ratio of 8 (f8) gives a focal length of 1600 mm (64"). The focal ratio of a telescope can be calculated by dividing the focal length of the primary lens or mirror by the aperture. For instance, if a 310 mm aperture telescope has a focal length of 1,920 mm, then it has an f/ratio of f6, (1920 ÷ 310 = 62). Amateur telescopes typically have f/ratios between f4 to f15. Telescopes that have focal ratios that are small, such as f4, mean that the focal length is very short and that the magnification of the primary mirror is only 4 times. A telescope with an f8 ratio will be twice as long and the primary mirror will magnify 8 times. Large professional telescopes can have very short focal ratios as low as f2 in order to keep the magnification down, so the field of view is wide. The shorter the f/ratio, the harder it is to figure the mirror to perfection, and the more likely it is to produce imperfect images if the optics are not perfectly collimated. Longer focal lengths from f8 to f15 are much more likely to produce sharper images and require less accurate collimation. They are good for high magnification planetary observations but not for wide field, deep sky objects. The longer the focal length, the longer the tube will be, so a ladder is usually required to reach the eyepiece when the telescope is pointed at objects high in the sky. Long telescopes can be overcome by folding the optical path back on itself as is done with the Schmidt-Cassegrain design.

MAGNIFICATION

Sometimes you will require low magnification to see large objects like star clusters or nebulas while at other times you need high magnification to observe small objects like the planets. You can change the magnification by changing the eyepiece. Eyepiece focal lengths range from 40 mm for low magnification, wide field views, to 2.5 mm for high magnification, narrow field views. There are other sizes of eyepieces between these extremes with focal lengths such as 13 mm, 9 mm, and 6 mm. These give a range of medium magnifications.

Magnification is written as a number with an 'x' after it e.g. 200x. To work out the magnification that a particular eyepiece gives on a particular telescope, divide the focal length of the main lens or mirror by the focal length of the eyepiece. A 150 mm aperture f8 telescope using a 12.5 mm eyepiece will have a magnification of 96 times i.e. 150 x 8 = 1,200 ÷ 12.5 = 96x. A 460 mm f4 telescope using a 4.5 mm eyepiece will have a magnification of 409x. High magnifications around 400x are usually the maximum, as atmospheric conditions and/or the quality of the main mirror seldom allow more.

Don't be impressed by a telescope's magnifying power. Even fairly small telescopes can magnify half as much as very large telescopes. My first telescope was a 60 mm refractor spotting telescope with only one magnification of 30x, yet as a young boy, I was so impressed to see Saturn's rings, Jupiter's moons, mountains and craters on the Moon, and bright star clusters and nebulas with it. I later purchased a Unitron 60 mm refractor which provided magnifications up to 300x. It delivered marvelous high magnification images under good seeing, however, using the 2 times Barlow lens which provided 600x, this did not resolve any more detail. Small telescopes usually magnify from around 30x to 300x and large amateur telescopes range from around 60x to 600x. It's best not to magnify your image to any more than twice your telescope's aperture in millimeters e.g. a 250 mm diameter mirror should not have a magnification above 500 times. Remember, very high magnifications also magnify any imperfections in the telescope's optics as well as poor seeing. You will find that the highest power you can use most of the time ranges between 200 and 350.

High Magnification

High magnification generally provides maximum resolution (clarity). However, if the seeing conditions are not good, then high magnification will just make a blurry image worse.

Medium to high power can be effective on very faint nebulous objects because it has the effect of darkening the background sky glow. This increases contrast between a faint object and the background sky thereby making the surface brightness of the object appear brighter.

Very small, faint objects can be below the eye's visibility threshold at low power because they are too small to turn on the minimum number of rods cells in the eye to make the object visible. High magnification can make a small, faint object like a planetary nebula, large enough to become visible. However, high power can also have the disadvantage of making very faint light become more spread out and therefore harder to see. But if the background sky light is reduced below that of the object, then the object becomes easier to see. You have to test this on each object. Give yourself time for your eyes to adapt to a darker field at high power. ⓒ I was very surprised when I noticed this occurring with the Pencil in the Vela supernova remnant. The Pencil Nebula is very faint and so thin that it was not visible at medium power in a 460 mm (18") telescope. However, when I used high power, this turned on enough rod cells in my eye for it to become visible with averted vision.

Throughout the book, reference is made to low, medium, and high power. For our purposes, these magnifications shall mean the following:

Very Low Power	= 30x to 60x
Low Power	= 65x to 120x
Medium Power	= 125x to 250x
High Power	= 255x to 400x
Very High Power	= 405x to 800x

EYEPIECES

Eyepieces provide a range of different magnifications.

There are many types of eyepieces and many different brands on the market. When planning to buy eyepieces, look up the brand and type on the Internet to see what experienced observers have to say about different types and brands. Those made by high-quality manufacturers that have fields of view over 50° wide, produce good sharp images when used on good quality telescopes.

The most impressive eyepieces are the **Nagler 82° wide angle** and the **Meade 82° Ultrawide Field** eyepieces. Although expensive, these brands have a very wide field of view. They make you feel like you are looking through a spaceship window! The great advantage of a wide field of view is that it allows the use of higher magnifications than you would otherwise use with a standard eyepiece. This produces more detail and more contrast while retaining a fairly wide apparent field of view. Low magnification, long focal length eyepieces in these brands are large and heavy. Because they have large lenses, they are more expensive than traditional eyepieces, but they are worth it for what they deliver. The less expensive **Pentax's SXW** eyepieces provide a 70° field of view with comfortable eye relief and improved contrast compared to narrow-field eyepieces. High-quality eyepieces range from US$300 to US$700 depending on the size and type.

Zoom eyepieces allow you to turn the eyepiece to increase or decrease the magnification. Typically, these types compromise optical quality by trying to marry a wide field of view at low power with a narrow field view at high power, however recent types have improved. If you only need to look at the center of the field of view, as when observing the planets, then they can work well enough. They do save time as you don't have to change eyepieces and refocus each time you change magnification.

Your eyepiece slides into a Barlow lens, like the one seen here, to increase the magnification of your eyepiece.

To save the cost of buying more eyepieces, a **Barlow lens** can be placed in front of your existing eyepieces to increase the magnification by 2 to 3 times, depending on the type of Barlow it is and how close it sits to the eyepiece. Cheap Barlow lenses that only have one lens element produce poor images with colored edges, so they are not worth using. Those with two elements work better.

USING AN OCCULTING BAR

Faint objects can be impossible to see if there is a bright object close to them. Some examples are the moons of Mars, Uranus, and Neptune. The planet swamps the faint light of their moons. Other examples are the Leo 1 dwarf galaxy that lies right beside the bright star Regulus or, the Veil Nebula that has a bright field star near its center.

 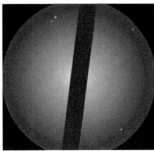

An eyepiece with an occulting bar

Left: Place an occulting bar made from aluminum foil, a strip of plastic or, opaque tape across the field stop of an eyepiece to mask a bright object. This will greatly assist in seeing an adjacent faint object.

Right: The brilliant glare of Mars can be hidden behind an occulting bar so its moons can be observed.

To overcome glare from a bright object, hide the offending bright object behind an occulting bar. To make one, fix a piece of thin, opaque card or tape across one-third of the eyepiece's field stop diaphragm. This is the black part with a hole in the center inside the open end of the eyepiece, as shown below. Alternatively, place a matchstick or two across the diagram's middle. You can then hide the bright object behind the occulting bar so the faint object can become visible. You will have to rotate the bar to achieve the best position for it so you can see the faint object. The occulting bar can be removed easily for general viewing. When bright objects are in a field, to reduce glare and a loss of contrast the optics of the telescope and the eyepiece need to be clean.

FILTERS FOR VISUAL OBSERVING

When used properly, filters can assist in making some astronomical objects and features more visible. All filters screw into the eyepiece except a sun filter, which goes over the open end of the telescope tube. Filters are often available cheaply on Internet sites that sell second-hand telescope equipment.

Moon filters are deep gray to reduce the glare of the Moon when viewed through a telescope. This stops you from walking around with a dark spot in your eye after looking away from the eyepiece. I find that most people like the extreme contrast that makes the lunar terrain so striking and they don't mind the blind spot in one eye for 10 seconds or so after viewing. When using high power, the Moon is not as brilliant as it is at lower power.

Colored filters can have a minor advantage in bringing out details on Mars and to some extent with Jupiter's gas belts. Martian frosts are a little more obvious with a blue filter and dust storms are brighter with a yellow filter. A green filter can emphasize the dark markings on Mars. Colored gels over the eyepiece will also work and save money but they are clumsy.

Colored filters for planetary observing **Lumicon Deep Sky filters**

Deep Sky Filters produce the most impressive results with emission nebulas because they effectively cut out unwanted artificial light pollution, as well as airglow. This greatly increases the contrast between an emission nebula and the background sky. The **Ultra High Contrast filter** and the **Oxygen III filter** do a superb job in this respect however, they cannot be used on star clusters, galaxies, or comets. Essentially, they only let through the green and red light of doubly ionized oxygen, which is abundant in most nebulas. They are great value at between US$100 and US$200. They are well worth this cost because, in effect, when observing nebulas, they make your telescope perform as if it had a much larger aperture. They do however hide faint stars in the field of a nebula, but this is a small disadvantage. The **Hydrogen Beta deep sky filter** does very little.

Glass Sun filters that fit onto the eyepiece of small telescopes provide great resolution but at a high price - your eyesight! They almost always eventually crack from the Sun's heat, so they are dangerous and should not be used. For this reason, they do not appear to be sold anymore.

A commercial aluminized Mylar solar filter fits over the open end of the telescope tube.

Aluminized Mylar solar filters go over the front end of the telescope. They are a safe way to observe the Sun because they reflect 99.99 percent of the Sun's light and heat. They diminish the full resolution capabilities of the telescope's aperture somewhat, but they are still a good way of seeing plenty of detail safely. They tint the Sun a bluish color. You can purchase aluminized Mylar solar film inexpensively and make your own frame to hold it. From the off-cuts, you can also make a small filter for your viewfinder, as you do not want unfiltered sunlight passing through it and burning you, or melting your viewfinder's crosshairs, or making the glue between lenses of its eyepiece boil and therefore making it useless.

Hydrogen Alpha Solar filters are built into special sun-viewing telescopes to provide marvelous views of the Sun that show its prominences, white flares, sunspots, and granulation. (See Volume 2 Chapter 3, page 79.)

TELESCOPE MOUNTS

Popular types of telescope mounts – top left - Alt-azimuth, top right - German Equatorial, Bottom left - Fork, Bottom right - alt-azimuth.

Commercial Dobsonian telescope mounts

Left: A manually operated Meade Lightbridge Dobsonian

Right: A Skywatcher computer controlled GOTO telescope

Less popular mountings are: top left - Split Ring, top right - English Yoke, bottom - English Cross Axis.

A telescope mount is an engineered structure that supports the telescope's tube. It enables the telescope to move around the sky.

Equatorial mountings for large amateur telescopes are costly compared to the invention of the **Dobsonian alt-azimuth mount**, popularized by the sidewalk amateur astronomer **John Dobson**. This mount dramatically lowered the cost of mounting large aperture telescopes. It has also made them more portable because a 'Dob' is easily disassembled into smaller parts. Because the construction of this design is simple, some amateurs build their own. Good quality, commercially-made Dobsonian designed mounts are inexpensive so there is less incentive to make one yourself - unless you have a passion for making telescopes, and quite a few amateur astronomers do.

WHAT TO LOOK FOR WHEN BUYING A TELESCOPE

When looking to buy a telescope, whether it is a new or used one, look for the following basic requirements.

An Equatorial mount (left) and a Dobsonian mount (right)

*The blue 310 mm (12.5") f5 telescope on the left utilizes an equatorial mount. The **polar axis** points to the celestial pole so that the telescope can follow the motion of the stars as the Earth rotates. The other axis, known as the **declination axis**, is at 90° to allow the telescope to move north and south. This telescope has rotating rings on the tube to allow the eyepiece to rotate easily to provide comfortable viewing positions. The timber 460 mm (18") f4.5 telescope to the right has a Dobsonian mount, also known as an Alt-Azimuth. The base rotates horizontally in the Azimuth axis by sliding on an inexpensive Teflon bearing. The tube moves in the vertical Altitude axis using a metal ring on each side that also slides on Teflon pads.*

1. **A good objective lens or mirror.** This is essential because without that, no matter what else the telescope has, it will be of little use. Nowadays, most telescope makers produce good quality optics, but this is not the case with cheap, flimsy telescopes that are sold in many department stores and in some camera shops.

2. **A good quality sturdy mount.** This is also a necessity. If you do not have a sturdy mount, the telescope will be unstable and very difficult to use. Cheap, flimsy,

equatorial mounts have small bearings that are not strong enough to stop the telescope from vibrating after you touch it, or from the slightest breeze. To test the sturdiness of a mount, tap the top of the telescope tube while looking into the eyepiece. See how quickly the vibration takes to die down. If it takes more than 3 seconds, the mount will not be sturdy enough to use. Cheap mounts also have cheap, plastic locking devices that break easily, and they have poor quality alloy bearings with slow motion gears that are easily stripped.

Be aware of cheap, flimsy telescopes.

Cheap telescopes made of plastic with flimsy mounts, poor quality eyepieces and viewfinders are close to useless. The marketing material promoting these telescopes often makes exaggerated claims regarding their performance.

3. **Good quality eyepieces are a must.** A good telescope can be greatly compromised by cheap eyepieces that have very small fields of view and/or blurry edges. Manufactures of cheap telescopes provide low-quality eyepieces to keep the total cost down. You need eyepieces with at least a 50° field of view, and preferably wider. If that means buying two extra eyepieces, then this would be a wise step because you don't want to compromise the functionality of the whole telescope by using poor quality eyepieces. Using cheap eyepieces can be likened to buying a good quality car and running it on bicycle wheels. (See further details about eyepieces on page 147.)

4. **A good practical finderscope** is needed if you cannot afford a computerized (GOTO) telescope that will find objects automatically for you. Too many viewfinder scopes are made of cheap plastic that often breaks after little use. Some are so cheap that they are just a hole to look through!

A finder scope needs to be a small telescope with a magnification of 4x to 7x with an objective lens aperture of no less than 25 mm but preferably over 35 mm so you can see faint stars in order to track down deep sky objects. The optics of an old pair of 7x50 binoculars make a very good finder when fitted to some tubing. Poor quality finder scopes have small apertures that give dark and/or blurry images and they have small fields of view. These factors make them nearly impossible to use.

The viewfinder needs to be solidly made and have good lenses and crosshairs. It also needs to have adjusting screws that make it easy to line it up with your field of view in your telescope eyepiece. The object of interest should be at the center of the viewfinder's crosshairs when it is in the center of the telescope's eyepiece. If you do not have a GOTO telescope, a poor finder can frustrate you to the point that you will not use the telescope. If the rest of the telescope is good, you can always buy a better quality viewfinder to replace a poor quality one. When buying a telescope, look through the viewfinder to see how practical it will be to use before buying the telescope.

A **Telrad viewfinder** can be useful as an additional *naked eye* viewfinder. It does not replace a telescopic view finder. It works by projecting red circles on a *naked eye* view of the sky. You use both eyes: one eye looks at the sky while the other eye looks through the Telrad. With both eyes open, the red circles in the Telrad appear to be projected onto the sky. A variable brightness control allows the Telrad's circles to be dimmed so that they are just bright enough to see so you do not compromise your dark adaption vision. I find it hard to get the circles dim enough when looking for faint objects. Place a bright naked eye star at the center of a low power eyepiece. Turn the three adjusting screws at the back of the Telrad so that its illuminated concentric circles are centered on the bright star. The Telrad's three circles indicate the field of view in the eyepiece of the telescope at different magnifications. The outer field circle indicates the size of the field of a low-powered eyepiece, while the tiny inner circle indicates the field size of a high magnification eyepiece. If you are looking for faint objects below naked eye visibility, then you will need to use your telescopic viewfinder.

A Telrad naked eye viewfinder projects red circles on the sky. The base plate is attached to the telescope's tube and the Telrad fits into this.

5. **A motor drive** is very helpful, because it allows the telescope to track the stars, so you don't have to keep moving the telescope by hand to keep up with the Earth's rotation. You need a motor drive that has large, solid metal gears that don't strip easily. The motor should run smoothly without changing speed or jerking when you are looking through the telescope. Most motor drives have a hand controller with adjustable speed control on both the polar and declination axes. This allows you to move around your field of view in any direction. The drive should be able to run off mains power, or a car battery if you take it on field trips. A good motor drive is worth every bit of the cost.

A GOTO computerized is not essential, but it is a great advantage because its computer control system will find any object for you and place it at the center of your eyepiece. You simply select the object you are interested in observing and the telescope will slew to it. A computer-controlled GOTO telescope can have either its own stand-alone computer, which can be accessed by a hand controller, or it may be connected to an external computer which provides extra functions. A GOTO telescope is initially set up by pointing it at a couple of bright stars like Sirius and Canopus, so it can determine its location on the Earth. Once an object is keyed in, the telescope will move to it within seconds.

Computer controlled telescopes automatically acquire objects.

Left: A commercial computer-aided telescope with a hand-controller.

Right: This commercial computer controlled telescope is connected to a number of external computers for conducting research. Credit: Greg Bock

Modern computers using networking technology allow photographic telescopes at remote observatories located in dark skies to be fully automated and controlled remotely from anywhere in the world. An astronomer at a remote location anywhere in the world can instruct the observatory to open the roll-off roof or the dome shutters, checking first that the weather monitor says it' is safe to do so. It then instructs the telescope to find the first object automatically that is on a list of objects for the night's observing run. The program instructs the telescope's camera to photograph each object of interest for a specified exposure time with specified filters. Image data collected by the telescope can be stored locally and accessed remotely by the observer at any time. In the event of cloud, rain, or wind at the observatory site, sensors will automatically shut down the telescope and close the observatory if need be.

An automated home observatory that houses a Meade 14" LX200 ACF telescope is located adjacent to an office control room. Credit: Greg Bock

A VIBRATION FREE MOUNT

A stable telescope mount is absolutely necessary to stop the telescope vibrating. Good quality commercial telescopes provide sturdy tripod mounts. To make these more stable, suspend a heavy weight or a bag of sand to the underside of the eyepiece tray. This greatly reduces vibration from wind and bumps. A solid mount will be compromised if it is placed on a springy, suspended timber deck that bounces a little every time you move. Many suspended decks today in modern apartment buildings are well engineered and built of concrete and steel so they are very stable.

If you are going to have your telescope in an observatory, then you are much better off making your own pier mount rather than using a tripod. A pier frees up space around the base of the telescope that would otherwise be taken up by the tripod legs. The pier can be constructed of steel reinforced concrete poured into a square plywood boxing, or easier still, a round PVC pipe 200 mm to 310 mm in diameter (8" to 12.5"). It's a good idea to have the pier expand out at the bottom into a larger conical footing so that it is less likely to move from vertical or sink. If your telescope is elevated in an observatory with a timber floor that is well above ground level, then the telescope will need a solid pier dug at least a 1.5 m or more into the ground.

IS BIGGER BETTER?

The most important thing that a telescope does is to gather a lot of light. Magnification is secondary to this. In the dark, the pupil in our eye opens up to 7 mm in diameter. By comparison, a 100 mm (4") aperture telescope gathers 200 times as much light as our dark-adapted eye, while a 310 mm (12.5") aperture telescope gathers over 2,000 times as much.

The larger the aperture, the brighter and sharper the image will be. Knowing this, it is common to catch 'aperture fever'. As you will often hear said, 'there's no substitute for aperture' or, 'bigger is better'. Very large telescopes can show a lot of detail, but only when the seeing is very good and, only if the mirror is well figured. The larger the mirror, the greater the chances that it may not be well-figured unless it is an expensive commercial brand. The advantages of brighter images and higher resolution with a large aperture telescope must be weighed up against it being much heavier and more cumbersome to move around. Large apertures are more easily affected by atmospheric conditions because they have a larger area for thermal air cells to move across the light path in the tube. This makes them more likely to suffer from bad seeing. If you want a large portable telescope with an aperture over 310 mm for field trips to dark sky sites, then it's best to be a Schmidt Cassegrain, or alternatively a Dobsonian, which can be disassembled into smaller parts for transportation. The resolution and light gathering capability of large apertures for deep sky observing will impress you, and so will the detail they show on Mars, Jupiter, and Saturn, but only when there is good seeing. Because many large homemade mirrors are not expertly figured to $\frac{1}{20}^{th}$ of a wave, they do not show any more detail than apertures half their diameter that have very good mirrors.

Left: **This steel-reinforced concrete pier on which a telescope is fitted extends through the floor of an observatory under construction.**
Center: **A tripod mount with fold away legs for transportation purposes.**
Right: **Celestron telescopes have well designed sturdy tripod mounts with adjustable legs.**

A giant amateur-built Dobsonian telescope

Dr. Erhard Hänssgen in Germany built this impressive 1070 mm (42") f/4.5 Dobsonian which can be dissembled to fit into a trailer, so it can be transported to a dark sky site. It provides plenty of exercise when climbing up and down the tall ladder required to get to the eyepiece. When moving from one object to another, the ladder has to be moved most times to get the eyepiece in just the right position, but the views make it worth the effort. Credit: E Hanssgen

A MOTORIZED OBSERVING CHAIR WITH A POWER-ASSISTED ELEVATING SEAT

Many amateurs are now buying or building very large aperture Dobsonian telescopes. These have eyepieces that are high above the ground when looking at objects high in the sky. This creates a market for a motorized observing chair; one that could elevate the observer to the required height of the eyepiece by using a joystick. As observers like to share what they are seeing with others, it would be a good idea for the chair to be a double seat that can move sideways so two people could take turns at looking at an object. Because such a chair would be heavy and likely to be used on grass, it would need wide inflatable wheels like those on pensioner's motorized buggies. The wheels would need to be steerable and electrically powered by a battery. The observer would need a hand-held paddle to operate it, so he does not have to come down each time he needs to move it a little. The wheelbase would need to be wide and take into account that the seat would need to get close to the telescope for objects close to the zenith.

SIMPLE OBSERVER'S SEATS

Standing for hours at the telescope can become tedious, so a chair and table that has an adjustable height makes observing at the eyepiece much more comfortable and practical. Remember the old rule: the more comfortable you are, the longer you are likely to want to observe, and the more you will enjoy it.

Because the telescope eyepiece often changes height when moving from one object to another, you need an adjustable height chair for visual observing. The larger the telescope, the more likely it is that you'll need to make a specially designed chair and/or footstool that has many variations in height to allow you to look into the eyepiece comfortably without bending or stretching.

It is also most advantageous to have a small swinging tabletop with an adjustable height so that it can be moved close to the eyepiece. This can be built into the observing chair. This is handy for holding eyepieces, a torch, star maps, photos, and drawing gear. Ladders are too easily toppled and the changes in height between the steps are too large. You do not want to be bending over or stretching upwards on your toes, as this will tire you very quickly.

CHOOSING A GOOD SITE TO SET UP

Observing chair options

Left: An adjustable height homemade observer's chair with a height-adjustable footstep and seat. The seat can also act as a small table top when standing. It can have a fold-out top to give it a larger area.

Center: These variable height chairs are commercially made. They are useful for refractors and Schmidt-Cassegrain telescopes where the eyepiece is at the bottom of the telescope, but they are not of any use with Newtonian reflectors where the eyepiece is at head height or above.

Right: The adjustable height 'Summit' observer's chair has built-in steps to get to the seat when it is in a high position. The steps also provide support for your legs when sitting. Such a chair could be built to be much higher for very large telescopes. It would need a broader base to make it stable and it would require wheels that are lockable. Credit: Diginow

It's best *not* to set up on a large area of concrete or bitumen that has had the full sun on it, because it will retain heat and produce poor seeing for an hour or more after dark. If it is in shade in the afternoon, retained heat will not be as much of a problem.

If you have your telescope set up for tracking the stars but you take it inside after each observing session, then it's a good idea to place it back in exactly the same spot the next time you use it. You can do this by placing some marks or hollows on the ground for the point of each leg to sit in.

To maximize the use of your telescope, read the next chapter to learn how to gain observational experience.

In the next chapter, we look at how to draw astronomical objects well, and how this greatly enhances one's observational abilities.

To see all types of commercial and homemade telescopes, and have a chance to look through them, go to astro-camps organized by astronomy clubs such as the Texas Star Party seen here.

Chapter 10
ENHANCING ONE'S OBSERVATIONAL EXPERIENCE BY DRAWING ASTRONOMICAL OBJECTS

A pencil drawing made at the eyepiece by the author

Left to right: Jupiter, Comet Bronsen-Metcalf, the M5 globular star cluster and the Jewell Box open star cluster.

WHY MAKE DRAWINGS?

Most people who become interested in astronomy are rewarded with seeing many impressive sights. It can be very satisfying to make a visual record of your observations because the ability to remember details fades very quickly, even within minutes, and over years it is almost impossible to remember anything more than a vague impression of the object, if that. Good drawings are invaluable as an aid to memory and they are far superior to written descriptions. As the saying goes: 'a picture is worth a thousand words'.

When I first started observing deep sky objects and comets, I would make written descriptions of my observations, but I soon found that many of them sounded much the same unless the descriptions were long, but then they were tedious to read. Even then, it was difficult to clearly visualize the object, so I decided to draw as many objects as I could so that I could remember clearly what each one looked like. After making a drawing, I recorded details of the observation such as the time of night, the date, the observing site, the weather conditions, my telescope details, how fresh I was, and whether the drawing was rushed. I found that because this information was added to the drawing, it would often bring back vivid memories of the many enjoyable observing sessions I had when making each drawing. This was especially the case if I was observing with someone else at the time, so I noted their presence.

Rather than make a quick sketch, I took a little more time, whenever possible, to make them reasonably accurate so I would have a more precise record of the characteristics of each object. Most drawings typically took me between 5 to 20 minutes, depending on how much detail was involved and how photographically accurate I made them. I found this small investment of time was well worth it as it allowed me to look back through my drawings to remember what I observed. For instance, when I reviewed all the drawings I had made of these objects I could immediately see the differences between one globular cluster and others. Drawings also show what can be seen using different apertures and different magnifications. You can also compare your drawings with those made by other observers.

COMPARING VISUAL OBSERVING TO PHOTOGRAPHY

Photographs do not show what you see visually. There is a considerable difference between what the eye sees compared to what photographs record. When you observe stars visually in a telescope, you see the stars as true pinpoints of light, regardless of their brightness. But photographs show stars as discs of different sizes relative to their brightness. In this regard, photographs are not as realistic. This is not to say that there is not great value in photographs because there certainly is. Digital photographs often show more faint extensions and more detail in nebulas and galaxies than the eye can see, but the brightest regions in photographs can be overexposed thereby masking detail that the eye can see. Photography can capture much more color than the eye can detect, but the eye can tell the difference between a star and nebulous cluster of stars in a galaxy's core, whereas with a photograph, this is not obvious. Visual observing and photography deliver different types of advantages, so they complement one another. Most photographs are wide-field, so they cannot reveal as much detail as high magnification on a telescope can.

A visual observation allows you to see sights and events in *real-time,* whereas with photography, you typically only see the detail of what you have captured sometime afterward when you are processing your images.

Observing astronomical objects or phenomena with your naked eye, or by looking into your telescope's eyepiece, is quite a different experience to doing photography where you sit idly beside your telescope while it is photographing, and later-on you get to look at the image on a computer screen. Understandably, visual observing makes you feel far more connected to the real universe than you do when you look at a photograph.

To set up for astrophotography can be a costly and time-consuming exercise, whereas visual observing and sketching can be done at virtually no cost and in minutes. To produce high-quality astrophotography, you need considerable technical experience and time to set up the telescope for photography. You'll also need plenty of time to learn to use digital processing programs. To become proficient at this can take years and it can cost many thousands of dollars. Many serious astrophotographers need to house their large, expensive photographic telescopes and equipment in an observatory, so this is a considerable additional expense in money and time. However, the learning curve for visual astronomy and making drawings is very much faster than that required to take high-quality photographs, and it is at no cost other than the telescope.

HOW TO SEE MAXIMUM DETAIL

Making drawings is an excellent way to learn how to see maximum detail visually. You certainly don't need to be an artist to have sketching sharpen your observing skills however, artistic talent will certainly help you to produce better quality sketches.

Making drawings teaches you the importance of taking your time to make good observations. Taking your time allows you to see all the details that are visible, as opposed to just looking briefly and thereby missing most of them. This is what almost all beginners do. When you make a drawing, you have to keep checking the telescopic image to be able to draw what it looks like. **Constantly checking every bit of detail quickly teaches you how to see all the detail that is visible.**

To make the most of visual observing, it is important to learn how to use your eyes. (See Chapter 7.) Your mind has the ability to build up a sort of short term composite image when it combines glimpses of detail that you see here and there when the seeing is steady. You quickly put the detail into memory and immediately transfer it to your drawing before you forget it. You then do that with the next bit of detail and so on. You have to keep checking the telescopic image to be sure that you have it right. The finished drawing becomes a permanent record of your observation.

Learning how to see detail

Left: When observing Jupiter, inexperienced observers usually take short glances not knowing what detail to look for, so they typically see little of what is actually there.

Right: Experienced observers will patiently scrutinize each part of the image over 10 minutes or more to enable them to build up a detailed view. Drawing by Gregg Thompson

We all start out as novices. I remember my early views of Jupiter: all I could see were two faint, light brown bands on either side of its equator. This was because I didn't know what to look for, and I didn't know how to recognize poor seeing. Ⓒ A few months later after much dedicated observing and experimenting with different eyepieces and different weather conditions, the same 200 mm telescope showed a wealth of detail.

Cyclones and streamers in the gas belts were visible as well as the moons transiting the planet and casting their eclipse shadows on the globe.

The more observing experience you have, the more you will see. With practice, you will learn how to get the perfect focus, how to recognize the best observing times, and to know when the planet is at its closest, and therefore presenting the most detail. Importantly, you will learn how to get the most out of your eyes. Quality drawings won't happen on your first few attempts: you need to make a dozen or so drawings before you intuitively learn the tricks of using averted vision and looking for fine detail when the seeing is good. You will then learn how to draw subtle changes in brightness and detail in the objects that you are observing.

This drawing of Saturn was made by Paul Rhea on the 19 May 2009 when one of Saturn's moons, Rhea, and its shadow, was transiting the globe. It was made using a 2,033 mm telescope.

HOW TO MAKE A DRAWING

◉ If you are not used to drawing, it's best to start off simply by using your naked eye to sketch the changing positions of the planets in the sky relative to other stars over a few weeks. Then try drawing the detail you can see on the Moon with your naked eye. This is much more challenging than you might think. Because the Moon is so bright, you can use an unfiltered, white light to make your drawing. Saturn is a good planet to start making drawings of because it does not have as much detail as Jupiter and Mars when they are at their closest to Earth. The positions of the moons of Jupiter and Saturn will change by the hour as they orbit their parent planet, so try recording their changing positions. The tilt of Saturn's rings changes over the years so you will be able to record this over time. When you feel like you are becoming proficient, try drawing Jupiter. It's large and bright and its storm clouds are always changing shape. Thermal upwellings from below the cloud tops cause huge cyclones to form.

When you first try to draw Mars, it is best to do so as many times as possible over a couple of months, before and after opposition when it is at its largest and showing the most detail. The changing aspects of its icecaps, its markings, and sometimes its dust storms, are fascinating to draw each night.

◉ With both Jupiter and Mars, if you make more than one drawing on the same night, the changing positions of their features will show the planet rotating

When you feel like a greater challenge, try sketching some features on the Moon as you see them in your telescope. Telescopic lunar detail requires considerable artistic skill. There is so much to see on the Moon that you could spend a lifetime making detailed drawings of lunar features. With the Sun moving across the lunar sky, the lighting angles are continually changing, and so is the viewing angle due to the Moon wobbling back and forth sideways, and also up and down, as it orbits the Earth. (See libration in Volume 2 Chapter 5, page 126.)

TO MAKE A PENCIL DRAWING

To make a pencil sketch, use a paper with a smooth surface such as white photocopy paper. Don't use paper with any texture as that will cause rough spots on your drawing. A 2B pencil is the best for drawing stars as pinpoints, and a blunt 4B pencil is best to use on its side for smudging in diffuse objects. Use a good clean, pencil eraser to make corrections. You may need to shave one end of the eraser to a point to be able to erase fine parts of your sketch that are in error.

It's not practical to use white pencils or chalk on black paper because white pencils appear gray on black paper, and you can't get a fine point with them. And chalk smudges badly. You could use white paint, but making a painting takes a lot more time and it can't be done well under low light. The most practical medium to work quickly and accurately with at the telescope in the dark is pencil and paper.

A sketch of a faint object made at the eyepiece under dim light will be a bit rough, but you can always clean it up afterward under white light to make the stars rounder, and to even out smudged areas that represent nebulas. I give bright stars fine diffraction spikes afterward to denote their brightness rather than using large dots. This looks more realistic and more like what you see in the eyepiece. If you draw a feature on the Moon, you will need to draw it quickly because the shadows move quickly across it as the Sun rises or sets. To make scientifically accurate drawings, make photocopies of the A4 drawing sheet at the end of this chapter. The circle is used to represent your field of view. Record the details of your observation in the boxes provided below it. (See page 168 and page 176 below for details.)

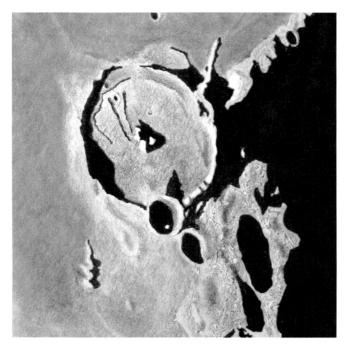 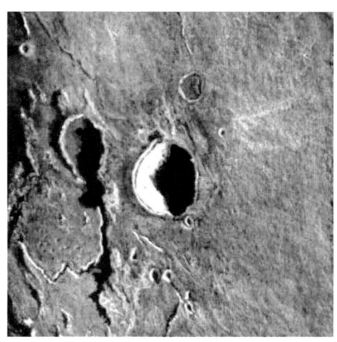

These excellent drawings of lunar craters were made through the eyepiece. They show very different drawing styles.

Left: This depicts the cracked floor of Gassendi drawn by Carlos Hernandez using a 230 mm (9") Maksutov-Cassegrain telescope at 359x under good seeing. North is at top and east is to the left.

Right: An expert painting in color of Aristarchus and Schroter's Valley by Alexander Massey

SKETCHING SUNSPOTS

By making drawings of sunspots, you can record their movement across the Sun's face, and also how they change shape hour by hour and day by day. One way to do this is to make a drawing as you see it in a medium to high-powered eyepiece. Another way is to project the Sun's image onto white card using a low to medium magnification. Draw a circle that represents the Sun's disc and center the projected image of the Sun on it. It's best to have this attached to your telescope so that the Sun does not move across your drawing paper due to the telescope following the Sun across the sky. By either means, sketch the size, shape, and position of sunspots each day that the Sun is visible over a week or more. You will see how sunspots develop and how they move across the face of the Sun as it rotates.

If you make some sketches of sunspots at high magnification, you will record much detail. When sunspots are near the limb of the Sun, they appear elongated and their centers can be seen to be depressed. This is due to seeing them being foreshortened due to being seen almost side-on. You will see considerable detail in their dark centers and their lacy, gray surroundings where flames are being drawn into the central depression. Sometimes brilliant white explosive flares are seen briefly at their centers. (See Volume 2 Chapter 3 page 72 to page 75.)

When the Sun was very active during its Solar Maximum between 1989 and 1992, it had many huge naked eye sunspot groups up to 20 times the size of the Earth! The following solar maximum from 1999 - 2003 was also active with huge sunspots erupting.

The maximum after that between 2012 to 2015 was much less active with very few sunspots of any size being visible with almost no large naked eye groups visible. It appears that the Sun will remain relatively dormant for some decades as it has in the past when it went into this phase. (See Volume 3 Chapter 3, page 68 onwards.)

This drawing of a medium sized sunspot was made using eyepiece projection onto white card using a 460 mm (18") telescope. It shows a short-lived, brilliant white flare erupting at the center of the sunspot. Credit: Gregg Thompson

SKETCHING THE PLANETS

To make detailed sketches of the planets, it's a good idea to have a motorized telescope to keep the planet right at the center of the field of view so you don't waste time having to track them manually to compensate for the Earth rotating. Use the highest power that gives the best detail and contrast for the seeing conditions. By the time you have made six or eight drawings, you'll notice that your sketches will have improved and that you are learning how to extract maximum detail by taking your time. Be subtle with your shading; don't make your drawings too contrasty.

DRAWING ROTATING PLANETS

The planets Jupiter, Saturn, and Mars are the best to draw as they display the most detail. Due to the rotation of their globes, and their changing weather, their detail is constantly changing. Saturn and Jupiter rotate fast in around 10 hours, so after only 10 minutes you will notice that features at the equator are quickly rotating away towards the planet's eastern limb. Those on the western limb are moving towards the center of the planet. With Jupiter, you can't waste time when you are making drawings of the features otherwise they will look distorted due to the planet's fast rotation. It's best to draw the detail on the side that is leaving the disc before you lose it, and then *roughly* sketch in the main features on the side of the planet that is rotating towards the center. Then draw the details at the center of the planet's disc. Once you have the broad outlines of the major features, you can then fill in the details. Use an eraser where necessary to correct mistakes, and to remove pencil to show bright areas. Limb darkening on Mars, Jupiter, and Saturn can be replicated by smudging the edge. The positions of Jupiter and Saturn's moons will change constantly as they orbit each planet so you can make a detailed drawing of that.

Venus and Mercury show only phases like the Moon. This is all you will see, as they reveal no detail, unless you have a large aperture telescope and can observe Venus during the day when its high in the sky. Sometimes subtle shading variation in the clouds along its terminator are faintly visible. Uranus and Neptune show only very small, dull discs so these planets are uninteresting to draw.

PREPARING AN OUTLINE FOR DRAWING THE PLANETS

Before you start making a drawing of a planet, you need to prepare an outline of it on some white card or paper. The outline for Mars should be 50 mm (2") across drawn on a white card about 150 mm x 100 mm (6" x 4") in size to allow room for your drawing details down one side. The discs for Jupiter and Saturn need to be oblate, not circular like the one for Mars. To achieve this, draw a circle lightly in pencil and then draw in the planet's outline on the outside edge of the circle where the equatorial region is and just inside the circle for the polar regions. Unless you are good at freehand drawing, so you can draw Saturn's rings of equal proportions on each side, and at the right angle, you will need to use a flexible French curve or plastic curves to draw them. Keep in mind that each year they will be either opening or closing depending on where Saturn is in its orbit. (See Volume 2 Chapter 10, page 357.) There needs to be room outside the disc for the rings. Saturn's disc can be smaller as it does not show as much detail as Jupiter. The moons of Jupiter and Saturn can be drawn in as small discs. The sky outside the planet's outline can be inked in with black ink.

The Ultimate Guide to STARGAZING Gregg Thompson

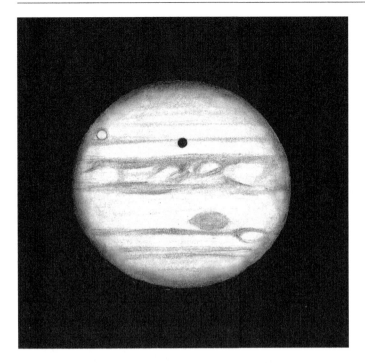

Photocopy these disc outlines of Jupiter and Saturn to make a drawing of the planets. Note that they are oblate.

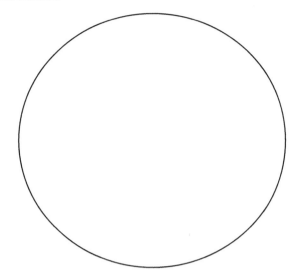

Outline for drawings of Jupiter

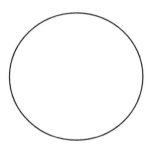

Outline for drawings of Saturn

These color drawings of Jupiter, Saturn and Mars were made by the author at the eyepiece of a 200 mm (8") reflector. Subtle colors have been enhanced.

Top: Jupiter's gas belts that continually change their structure. Note the moon Callisto and its shadow transiting Jupiter.

Middle: Sometimes storms erupt in Saturn's cloud belts. They appear as disturbances along the edge of the equatorial belt, as seen here. Note the divisions in the rings, the shadow of the rings on the globe, and the shadow of the globe on the far side of the rings, as well as four of Saturn's moons.

Bottom: Mars' icecaps change shape from summer to winter and some of its markings change in darkness. About every 15 years, Mars has a planet-wide yellow dust storm that hides all detail. Note the division in Mars' icecap and its darkest feature below center known as Syrtis Major. Its dark color is caused by the land there being a low-relief shield volcano that has left a large, high altitude plateau that has little yellow dust over it. The rock is dark as it is volcanic basalt.

OBSERVING MARS

If you draw Mars from a couple of months before opposition to a couple of months after that, you will see the greatest amount of detail. Outside of this window of time, Mars becomes so small that it is hard to see much detail.

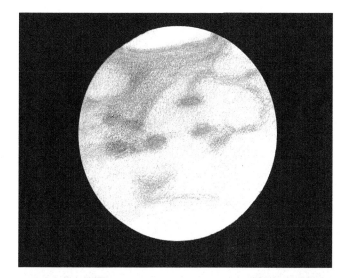

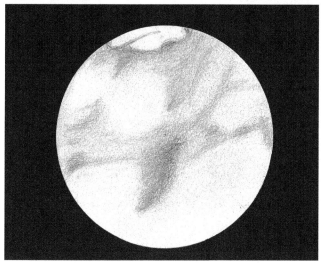

Two views of Mars

These drawings show opposite sides of Mars. Earth revolves around the Sun in almost half the time that Mars does, so before and after Earth's closest approach to Mars, we see it a little side-on. This produces a ¾ phase effect as seen in the drawing on the left. Drawings by Gregg Thompson.

Opposition occurs every two years. Ⓒ The changing aspects of Mars' icecaps, its dark markings, and the development of dust storms make interesting drawings of over a couple of months.

This high-quality drawing of Mars was made by a professional artist, Laurie Hatch. She had no experience in planetary observing when she was given the opportunity of drawing Mars through the large 36" (915 mm) refractor telescope at Lick Observatory – probably the best telescope in the world for visual planetary observations. Laurie did not use any photographic references: she drew what she saw in the eyepiece. To make a drawing this good on the first go with no prior experience of what Mars looks like, and having to allow for air turbulence and how Mars rotates as you are drawing it, is quite amazing.

Laurie used a 250mm (10") bowl to draw the outline circle. She sanded her drawing paper to get a smooth surface. To reproduce Mars' pink color, she used a chamois dipped in dry ferrous dust rust and lightly stroked and rubbed the surface of the paper to get a uniform color. An eraser removed areas that needed to be lightened. A soft 4B pencil was used for the darker regions. Laurie said that this rare evening was one the most memorable and satisfying in her life.

Ⓒ Mars rotates 40 minutes slower than Earth so if you observe it at the same time each night, over 36 days you will end up seeing the whole surface passing across the center of the planet.

OBSERVING JUPITER

Jupiter's globe is large and bright and its ever-changing cloud bands show many **hurricanes and cyclone**s. Thin **streamers** are often drawn out of a belt into the adjoining band. Jupiter displays considerable **limb darkening**. This is caused by the planet's thick atmosphere absorbing sunlight being absorbed along its limb and not reflecting back to us. The polar regions display little detail.

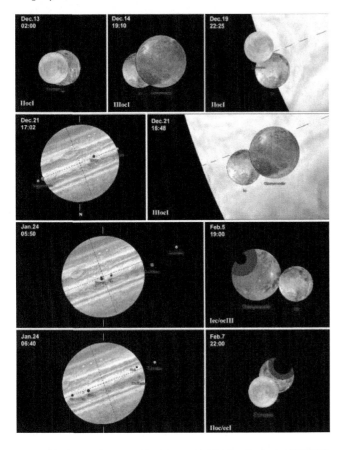

Jupiter's four largest moons revolve around it at different rates. As they pass in front of the globe, they cast their **eclipse shadows** onto Jupiter's cloud tops. Predictions for these frequent events are available on the Internet. Transits of moons and their shadows make interesting subjects to draw. Every six years, the **Mutual Phenomena of Jupiter's Galilean moons** occur whereby one moon occults another or, one moon casts its shadow on another. Predictions for when these events occur are also found on the Internet. When these rare events are viewed using high magnification under good seeing in a telescope of 250 mm or larger, they are opportunities to make very interesting, although difficult, drawings. I have observed these phenomena but not drawn it.

Ⓒ Amateurs can draw the mutual phenomena of Jupiter's satellites. To my knowledge, this has not been done.

Drawings of parts of a disintegrated Shoemaker-Levy comet/asteroid in 1994 that impacted Jupiter's upper atmosphere left dark dust clouds. The largest dust clouds were larger than the size of the Earth! Such obvious 'scars' were totally unexpected. These drawings were made with a 310 mm reflector by the author.

OBSERVING SATURN

When observing Saturn, you'll easily see its gas belts, rings, and moons. Every 14 years, huge storms will develop in the temperate gas belts. (See Volume 2 Chapter 10, page 359.) The tilt of the planet's axis makes the angle of one pole lean towards us for seven years to varying degrees, and for the next 7 years, the other pole will do the same. Between these times, the rings are seen edge-on. They become so thin for a day or two that they become invisible. (See Volume 2 Chapter 10, page 358.)

When the rings are close to edge-on, this is when Earth passes through a line between Saturn and the Sun. At this time, you will see the **Mutual Phenomena of Saturn's moons** occurring. This is the only time that the moons pass in front of Saturn's globe and cast their shadows onto its cloud tops. (See Volume 2 Chapter 10, page 358.)

Titan's shadow is visible in telescopes with apertures of 250 mm (10") and larger. The discs and shadows of the other moons are too small to be seen in amateur telescopes. The moons only transit Saturn's globe for a short time due to the plane of the Moons being aligned with Saturn's axis, which is tilted at 27° to the orbits of the other planets. By comparison, Jupiter's axis is tilted at only 3° so its moons cast shadows on its much larger globe all the time.

HOW TO DRAW DEEP SKY OBJECTS

If you enjoy drawing, you may wish to draw many types of deep sky objects. I have made hundreds of drawings of almost all nebulas and star clusters in both the Milky Way and the Magellanic Clouds. This took many years to accomplish. Some of these drawings are shown in this book.

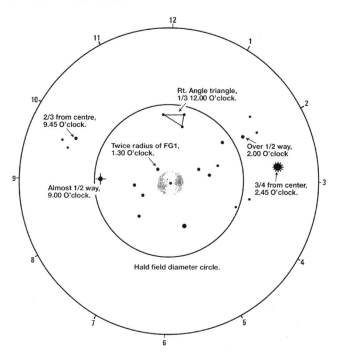

How to achieve good proportionality of field stars

When drawing deep sky objects, choose a magnification that best fits the object inside the field of view. If it is a small planetary nebula or small distant galaxy, then use the highest power the conditions will allow. If it is a large object, then use a low power to have it all in one field. For large bright emission nebulas such as Eta Carina or M8, you may like to make a second detailed drawing at high power of a small bright section.

ACHIEVING GOOD PROPORTIONALITY

To draw the proportions of the object and its star field accurately, think of the field of view as a clock face and then divide it into four or eight pie-shaped segments. If there are more than a few stars in your field of view, then divide the diameter in two by having a faint circle with a diameter half that of the outside of the field. You can use a compass to make this circle and rub it out when you are finished.

Should you want to achieve near-perfect accuracy in the positioning of stars in your star field, find a photograph on the web or in a book of the object you want to draw, and make a photocopy of it at the scale you want your drawing to be. Place this under your drawing sheet before you start, and lightly mark the positions of the brightest stars. When drawing at the eyepiece, you can draw in these stars to the brightness you see them visually. Keep in mind that the brightness of stars in photographs will seldom be the same as they appear visually. This is because stars that are very blue or very red will be seen to have a different brightness to the eye to those recorded on a camera's CCD. This is due to CCDs having a more even sensitivity to color compared to the eye. However, the eye detects subtle differences in the brightness of stars more clearly than a camera can. This makes magnitude differences in drawings more pronounced than in photographs.

To start your drawing, determine where north is by moving your telescope slightly to the north. Place the object you are drawing close to the center of the field. Your telescope's drive should keep it centered. In each quadrant, place the brightest stars as accurately as you can. Determine where the bright stars lie on an imaginary clock face (e.g. at 2 o'clock) and then note how far each star is from the center to the edge e.g. half way or two thirds etc. Use the bright stars as base points for placing fainter ones that are near them. Note the angles and distances between them. You will see groups of stars as shapes e.g. lines, triangles, curves, squares, parallelograms, pentagons etc. This makes it easier to place them. Some deep sky objects will have very few field stars involved, especially if they lie well away from the Milky Way, or because you are using high magnification. However, objects that lie in rich Milky Way star fields will have many faint stars, especially when using low or medium magnifications. Unless you are drawing star clusters, it is not as important to achieve high accuracy for the outlying field stars as it is for the object you want to draw. But if you have the time, then try to position all the stars accurately.

Drawing Nebulous Objects

When drawing deep sky objects, you will be working at the limit of your vision, so you will need maximum dark adaption. To achieve this, use as little light as possible on your drawing and ensure there is no stray light entering your eyes. For the very faintest objects, it's a good idea to use a hood around your eyepiece to maximize your dark adaption and averted vision. (See Chapter 7, page 117.)

To draw nebulous objects such as comets, diffuse nebulas, planetary nebulas, and galaxies, use the side of your lead pencil to rough out the general shape of the object softly and then start to build up density where the brightest regions are. Gradually get the right proportions in the same way that you would when plotting field stars.

Remember that making a fairly accurate sketch will force you to scrutinize the object to extract the maximum detail your eye can detect. After a while, you will do this intuitively. Drawing objects forces you to become a very good observer, even if you are not a naturally great artist.I made detailed drawings of over 300 of the brightest galaxies and their field stars from enlargements I made from the 48" Schmidt plates. These drawings were made into A/4 charts that had a diameter of half adegree, which gave a scale of 30"/ mm. This was done so that amateur astronomers could visually discover supernovas in these galaxies.

To make the charts, I placed the Schmidt photographs on a lightbox over which I laid my visual observation sheet for each galaxy. This way, I could plot very accurately, the field stars that I could see visually. I placed the center of the galaxy at the center of the chart. I ensured that each star's magnitude was as it was seen visually - not as it was in the photograph. I then noted which bright spots in a galaxy appeared like field stars, but which were not visible. Under high magnification, they were shown to be large nebulas or bright open clusters. I rendered these as diffuse spots so they could not be confused for a supernova.

When the finished art for the charts was completed, they allowed amateur astronomers to place them on a dim lightbox to compare what they saw in their eyepiece with the chart. This way, any new star that appeared would be a supernova, if it was not identified as an asteroid. Amateurs were then able to discover supernovas visually soon after they started to explode. To estimate very accurately the changing brightness of supernovas as they started to rise to maximum and then fade away over the following weeks or months, I asked professional observatories around the world to provide me with whatever photoelectric measurements they had of field stars around galaxies. I then put these in the charts. This way, amateurs can quickly compare the brightness of a supernova to certain field stars with an accuracy of a fraction of a magnitude. Or, they can do so less accurately to half a magnitude by comparing the supernova to the size of the dots of field stars on the charts using the magnitude scale at the bottom of the chart. They can measure the offset of the supernova from the galaxy's nucleus quite accurately by using the chart's scale.

This work took 12 years. It was published by Cambridge University Press as '***The Supernova Search Charts and Handbook***'. It led to a substantial increase in the number of visual discoveries of supernovas. Because these discoveries were made *as a supernova was occurring*, this permitted professional astronomers using large telescopes to make detailed observations of them while they were on the rise and thereafter. Until then, discoveries of supernovas on the rise had been rare events that were mostly recorded in photographs by chance. Amateurs observing these galaxies each night were able to see a new supernova as soon as it rose above their threshold. Having real-time observations of these exploding stars permitted professional astronomers to observe them with large telescopes and take spectra of them. Astrophysicists could then develop more accurate models of the physics involved in different types of supernovas. This led to certain supernovas being used as 'standard candles' to measure distances to galaxies across the universe. Astronomers were then able to accurately determine the universe's size. Drawings made at the eyepiece can sometimes have much greater value than you might think.

Left: **M101 drawn by Jeremy Perez using a 200mm (8") Newtonian reflector at 120x**

Right: **A drawing by Cseh Viktor of M31 and its companion elliptical galaxies M110 an M32 at the very low power of 32x to see it all in one field**

Two sketches of NGC 5128 by two observers using different apertures.

Left: *NGC 5128 using a 710 mm (28") telescope using a range of magnifications, but mostly 438X. Credit: Serge Viellard*

Right: *NGC 5128 drawn by Peter Kiss using a 400 mm (16") telescope at 180x in Namibia's dark skies.*

Left: This excellent pencil sketch of a Markarian chain of galaxies using a 250 mm (10") telescope was drawn by Michael Vlasov.

Right: The galaxy M104, also known as the Sombrero Galaxy was made by Jeremy Perez using a 200mm (8") Newtonian reflector at 240x.

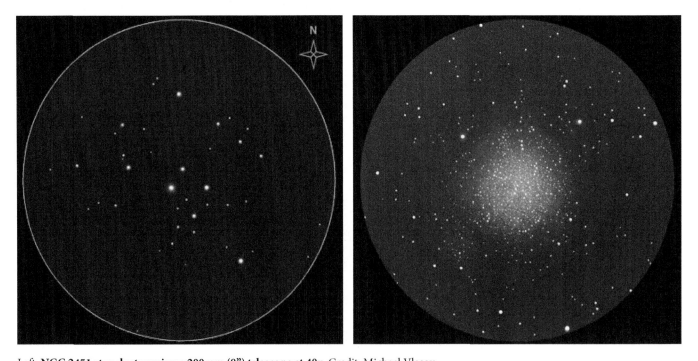

Left: **NGC 2451 star cluster using a 200mm (8") telescope at 40x.** Credit: Michael Vlasov

Right: **NGC 5139, Omega Centauri using a 260mm (10") telescope at 106x..** Credit: Michael Vlasov

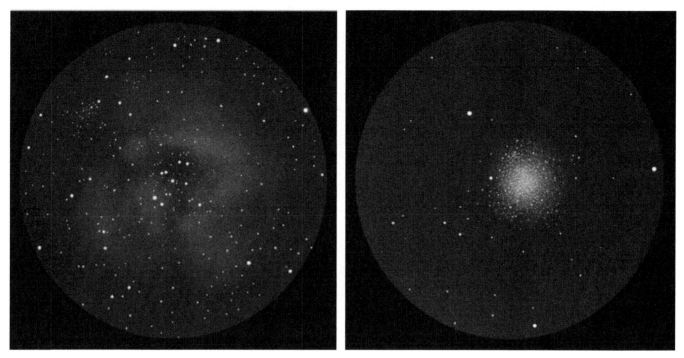

Left: **A very realistic drawing of the visual appearance of the Rosette Nebula using a 200mm (8") telescope at 32x.** Credit: Michael Vlasov

Right: **Messier 2 using a 260mm (10") telescope at 240x.** Credit: Michael Vlasov

This superb drawing of M51 drawn by Brian Banich using a 710 mm (28") telescope shows what a large aperture telescope will reveal. Brian used a range of magnifications but mostly 438x. To see the most stellar points he used 710 – 816x.

Serge Vieillard made this drawing f M31 with its companion elliptical galaxies using a 16" telescope at low power.

This Supernova Search Chart for M101, NGC 3031 is one of 300 prepared by the author to discover supernovas visually. Visual magnitudes for measured stars are given to assist observers in estimating the magnitude of a supernova. Within the circle, the field is half a degree. The nucleus of the galaxy is at the center. The offset of a supernova from the nucleus can be easily and accurately measured.

Deep sky drawings by the author that illustrate the visual appearance of the Ghost of Jupiter planetary nebula NGC 3243 at 660x (left), the Swan Nebula, M17 at 187x (center), and the head of Comet Halley at 60x on 12 March, 1986 (right). All drawings were made with a 310 mm reflecting telescope and then cropped. They were all drawn using pencil on white paper. They were later scanned in reverse so that they would appear as they do to the eye.

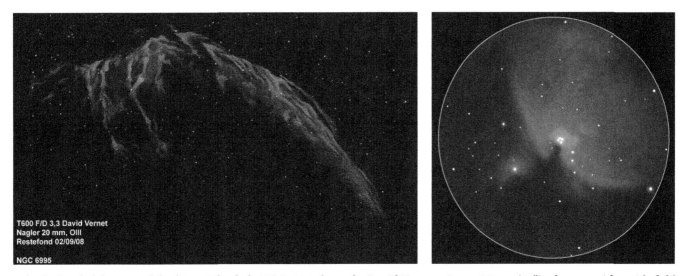

A finely detailed drawing of the faint Veil nebula NGC 6995, drawn by David Vernet using a 600 mm (24") telescope with a wide field 25 mm Nagler eyepiece and a OIII filter to enhance contrast. M42 (right) drawn through a 200 mm telescope by Michael Vlasov

HOW TO MAKE NEGATIVE VERSIONS OF SKETCHES

When you draw a deep sky object in pencil, you are recording a negative view to what your eye sees. To convert your pencil drawing on white paper to white stars on a black sky, use a photocopier to scan your drawings and then a photo-managing computer program to reverse the scanned image from a negative to a positive one. You can also use this method to make a negative color drawing to become a positive color drawing.

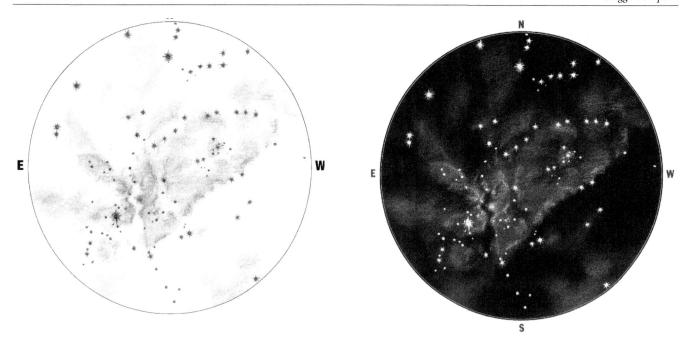

Left: **This is a pencil sketch of the Keyhole dark nebula at the center of the Eta Carina nebula. It was drawn using a 310 mm (12.5") telescope at a power of 176x with a wide field eyepiece.** Drawing by Gregg Thompson

Right: **The Eta Carinae Nebula is shown here as it was drawn with pencil on white paper (top) and the same drawing scanned and reversed using the Picasa photo managing program (bottom). Photoshop will do the same.**

HOW TO MAKE COLORED SKETCHES

You can use colored pencils to add color to your sketches. Colored pencils allow you to capture the subtle colors of the planets, double stars, bright giant stars in clusters, and the colors of some planetary nebulas and comets. When you begin observing, be aware that your eyes have not yet been trained to detect subtle color differences. Most objects will look colorless at first, but after a while, you become quite sensitive to subtle differences in color - just as you become more aware of subtle detail as you make more drawings.

You can add color to the planets and bright stars while you are drawing them. Bright objects can be drawn under dim white light with no dark adaption problems. However, to add color to a planetary nebula drawing, or those of comets, you have to add color to the finished drawing after it is completed because you will need bright light to do so. This means you will have to remember the color, or make a color patch using colored pencils straight after you make the drawing. Because you will be reversing your drawing to have white stars on a dark sky,

These pencils are arranged so that complementary colors are opposite one another. If you use the colored pencil opposite to the color you are seeing in the eyepiece, you will get the color that you want when you reverse your drawing.

then you will need to make the color of your deep sky object the reverse color to what you saw in the eyepiece. This way, when you reverse your drawing to take it from a negative to a positive, the color will be what you saw in the eyepiece. This works the same way that negative color film did when it was reversed during printing.

To know what colors to use to make your drawing, choose the colored pencil from the color chart below that is directly opposite the color you want your reversed drawing to be. For example, if a star is bright blue, then use a magenta (bright pink) pencil. If the color in the eyepiece is orange, then use a medium blue pencil. Sometimes you will need to blend two colors to get the right color.

Colors in the eyepiece are quite subtle but if you try to make a drawing that has that level of subtlety, then you will find that your drawing will look washed-out and not like it was in the eyepiece, so you will need to enhance the color and the contrast a little to make it look right. The eye can see far more steps in tone (grays) and many more colors than you can get using colored pencils. When you see direct light in the eyepiece, this allows your eye to see far more variations in color and brightness than you can see from reflected light off paper. Paper can only reflect eight differences in tone, whereas direct light has at least 32 shades that the eye can detect. It is the same for colors - the eye can see far more colors from direct light than it can see from reflected light.

How to Change a Negative Drawing to a Positive One

To reverse your negative drawing, scan it with a photocopier and select the reverse (or invert) setting. Or better still, you can send the scan to your computer, and by using a photo managing program like Picasa or Photoshop, you can reverse it. This will make it look like it does in the eyepiece. These photo managing programs will allow you to tweak the colors and also to get the right contrast or brightness you desire for the final product.

THE COLOR SENSITIVITY OF THE EYE

The eye sees color differently to a digital camera. For astronomical objects that have low surface brightness, the eye will see blue-green light more easily than red.

The very orange star Eta Carina lies at the heart of the large Eta Carina nebula. It is surrounded by a reddish-orange nebulosity known as the Homunculus nebula. The negative colored pencil drawing on the left was made by the author using a 310 mm (12.5") telescope at 660x. It was made in pencil and the star and nebula were colored turquoise after the drawing was made. Turquoise is the opposite color to the color it appears visually. To make the nebula appear orange on a black background, the original negative color drawing was reversed.

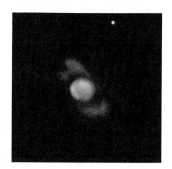

This is the small, bright turquoise planetary nebula NGC 6572 as seen with a 460 mm (18") f4.3 reflector at 754x. It was sketched in pencil and colored using the opposite color. To make it appear as it does to the eye, it was reversed to appear the color it is. Credit: Gregg Thompson

Left: The original color drawing produced in reverse color.

Right: After it is reversed, it appears turquoise as it does in the eyepiece.

These are reversed pencil drawings that were made by the author using 310mm reflector.

Left: A drawing of the strangely shaped planetary nebula NGC 5189 at 330x. It displays no color.

Right: The greenish-blue Saturn Nebula NGC 7009 at 176x

Interestingly, red stars become brighter the longer you look at them. For bright objects, our eyes see yellow-green more strongly because this is the most intense color in the spectrum of sunlight. We evolved to see best with this color. Modern digital camera chips have a flat color response, so they see all colors much more equally in low light than our eyes do. Digital photographic image-processing programs can greatly enhance colors in the original image and change them to achieve a result that is pleasing to the eye. But photographic colors are not seen by the human eye unless they are bright, and even then, the eye sees them a little differently.

Here we have a color drawing by the author of the beautiful, rich, and bright star cluster NGC 3766 in Carina.

This star cluster displays a very bright, blue-violet giant star (lower left) and an equally bright red giant (top). There are also a couple of less bright red giants near the center of the cluster. The original drawing has been cropped and reversed. It was made with a 310 mm reflector at 66x.

To most observers, the center of M42 looks slightly bluish, but some observers see it as being more greenish. The greenish appearance may be due to color blindness, which can eliminate blue light. A camera on a small amateur telescope will easily capture the brilliant reds or pinks in the inner parts of the nebula's shell as well as blues in the outer edges. The colors of most nebulas are too faint to be seen by the eye except when using very large apertures, and then most are only seen faintly. The eye needs a very large aperture telescope of 460 mm (18") or larger to detect just a hint of the pinkish-red color in the brightest parts of the outer regions of M42, the Orion Nebula. These parts appear more like a faint brownish tinge than pink. Camera images can easily be enhanced to bring out faint colors but there is no ability for the eye to do that. In time, genetic engineering and nanotechnology will allow the human eye to have a much greater sensitivity to faint light and color. In the future, today's night-vision goggles will be upgraded to become full-color night vision goggles.

If we could view the Orion nebula from close by in a spaceship window, we would not see it anywhere near as well as we do visually in our telescopes from Earth! This is because its gas would be spread out over a huge area making it far too faint to be visible to the naked eye. Even the very center would be difficult to see due to the glare of the bright Trapezium stars.

Visually, the center of the Orion Nebula is similar to the blue-green appearance here. This image was taken using a hydrogen alpha filter. Credit: Greg Bock

The human eye needs relatively bright light to see color. Ⓒ Even the bright light of a Full Moon is not enough to allow us to see color in our surroundings. If in the future, we were able to go sightseeing in the outer solar system, the rich colors of the gas giants and their moons that we see in space probe photographs, would appear colorless to our human eyes due to the Sun's light being so dull at that distance. We would have to wear goggles that intensify light and color to see the colors that exist on these worlds. The chances of humans ever visiting these worlds physically is extremely remote, so full color night goggles will not be necessary. But by using virtual reality, we will be able to visit other worlds instantly and safely. This way, we will see them as if they are lit by bright sunlight.

A typical picture of M42 showing its pinkish interior of fluorescing hydrogen gas. Its outer shell is bluish due to light from bright blue stars reflecting off the dust. To capture the faint outer regions with a camera, the much brighter center becomes overexposed and therefore appears white and burnt out. Credit: Greg Bock

By using virtual reality technology, we will not need a cumbersome space suit to lug around, and the risk of death from intense pressures and cold would be eliminated.

NIGHT SKY PAINTINGS

If you are artistic, you may like to paint a scene during a total eclipse of the Moon, or a total solar eclipse. Perhaps you could capture a scene of a brilliant meteor blazing brightly over a landscape. A very atmospheric art piece would be a painting of the Milky Way as seen under a dark country sky. Such paintings could include a campfire in the foreground with people sitting around it on logs. There may be the silhouette of nearby trees contrasting against a starry sky. To add atmosphere, you could have internally lit farmhouses in the distance, and mountains along the horizon silhouetting against the sky. Some high cloud may be drifting across a part of the Milky Way while a faint meteor streaks across the heavens above. Such a painting would be very rare, so it would be quite special.

Mike Smith painted this beautiful view of the Milky Way reflecting in a lake near the Mt Hood volcano in Oregon, USA.

This was a quick watercolor painting I made of Comet Kohoutek in the western twilight sky as it set behind a storm cloud over a mountain peak. The storm cloud was frequently being lit by internal lightning. This painting was made quickly to help me always remember how special that scene was.

RECORDING SCIENTIFICALLY VALUABLE OBSERVATIONS

To give your drawing more value, record the variables that affect your observations. This will allow you to compare old observations with past and future ones, as well as those done using different instruments. You'll also be able to compare them with those made by other observers. It takes less than a minute to rate the variables, so it is well worth doing this. There are three groups of variables that should be recorded:

1. **INSTRUMENT** - The details of the instrument used
2. **CONDITIONS** - The observing conditions
3. **OBSERVER** - The observer's details

These can be filled in at the base of your observation form. An A4 version of this can be found at the end of this chapter for you to photocopy.

Fill in the name of the object at the top and where applicable, the constellation in which it is found. Cross out whichever of the E/W makers is not relevant. You can easily determine what is the east side of the field by simply moving your telescope tube slightly in that direction. The same can be done to determine N/S. It should be straight forward to note the details of your telescope.

RECORDING OBSERVING CONDITIONS

Rate the observing conditions as given below.

Seeing: The term 'seeing' is used to describe the stillness of the atmosphere. (See Chapter 6, page 96.) **Good seeing** is especially necessary for observing fine astronomical detail. This usually occurs when a high-pressure system is overhead. An easy way to gauge the seeing conditions with the naked eye is to judge the extent that stars are twinkling i.e. 1 = z steady, 2 = occasional twinkling, 3 = rapid twinkling. A telescope will greatly magnify poor seeing that is seen with the naked eye. Advanced observers can measure the seeing in 'seconds of arc' e.g. moderate seeing might be 5 seconds of arc, meaning this is the finest detail that can be seen, whereas, in good seeing, it may be as small as 1 arc second or better.

Transparency: The perception of the brightness of nebulous objects depends on how much it contrasts against the background sky. If the transparency of the atmosphere is not good, then contrast is lost and very faint nebulosities become invisible. The clarity of the atmosphere is important for transparency, but the slightest haze or thin high cloud can be difficult to detect if you are at a dark site away from city lights. It's only when the object you are observing varies in brightness that you realize that thin cloud is slightly obscuring parts of the sky when you look carefully with your naked eye. Transparency can be badly affected by forest fires that leave smoke lingering in the air. Large volcanic eruptions leave wavy, high altitude bands of fine ash in the upper atmosphere for months and it can spread over thousands of kilometers. It is only seen at the end of twilight, but it typically has an imperceptible effect on transparency if one is not close to the source of the eruption.

Classify transparency as:

E = excellent, **G** = good, **F** = fair, **P** = poor, **B** = very bad

Sky Darkness: If the sky is not fully dark, then contrast is considerably reduced, and faint objects are invisible. This can occur due to twilight, moonlight, or city light pollution. The following ratings are useful to quantify sky darkness:

1 = perfectly dark, **2** = fairly dark, **3** = 3-4 day old Moon, **4** = 7 day old Moon, **5** = Full Moon

Wind: Wind can be a major deterrent to making good observations. When it's windy, the seeing is typically very bad and your telescope will vibrate causing detail to be lost. Wind often makes it cold and papers are hard to hold down, so you rush observations causing them to be compromised. If there is much more than a light breeze it is impracticable to make a drawing. Wind can be rated as:

W = gusts of wind, **B** = a light breeze, **N** = no wind.

Temperature: Cold temperatures make eyepieces dew up and observing becomes uncomfortable, so the quality of the observation is compromised. While the exact temperature can be recorded if you wish, it is the psychological effect that is more important because some people are affected more by cold or heat than others. Record it as it feels to you:

F = freezing, **C** = cold, **M** = mild, **W** = warm, **H** = hot.

Dew: Heavy dew can be a problem when you are observing out in the open. It can fog lenses and mirrors as well as make maps and your drawing paper damp. As this can severely compromise an observation, note it as:

H = heavy, **M** = mild, **N** = nil

Location of Site: This is good to record as it helps you remember where you were that night. Also, note the altitude and any other special particulars.

Recording Observer Variables

The observer variables that should be recorded are:

Visual Acuity: Note your visual acuity number by using the test for seeing detail provided in Chapter 7, page 119.

Threshold Vision: Note your threshold vision number for how faint an object you can see by using the test for this in Chapter 7, page 115.

Experience: Experience affects how much you will see so this should be noted as:

1 = expert with years of observing experience, **2** = somewhat experienced, **3** = Casual observer, **4** = novice.

Fatigue: Tiredness will affect the quality of your observation because it creates a tendency to cut corners and overlook detail or faint objects. Fatigue makes you become forgetful and lose concentration. It will change as the night wears on. Rate it as:

1 = fresh, **2** = a little weary, **3** = very tired.

Drawing accuracy: The accuracy of your drawing should be noted as follows:

1 = Photographic-like accuracy, **2** = Object accurate but field stars roughly placed, **3** = Object fairly rough, **4** = a rough, quick sketch.

You can also make a note of other people who were observing with you. An A4 Observation Form follows to print and use at the telescope.

In the next chapter, we look at how to take exciting astrophotography.

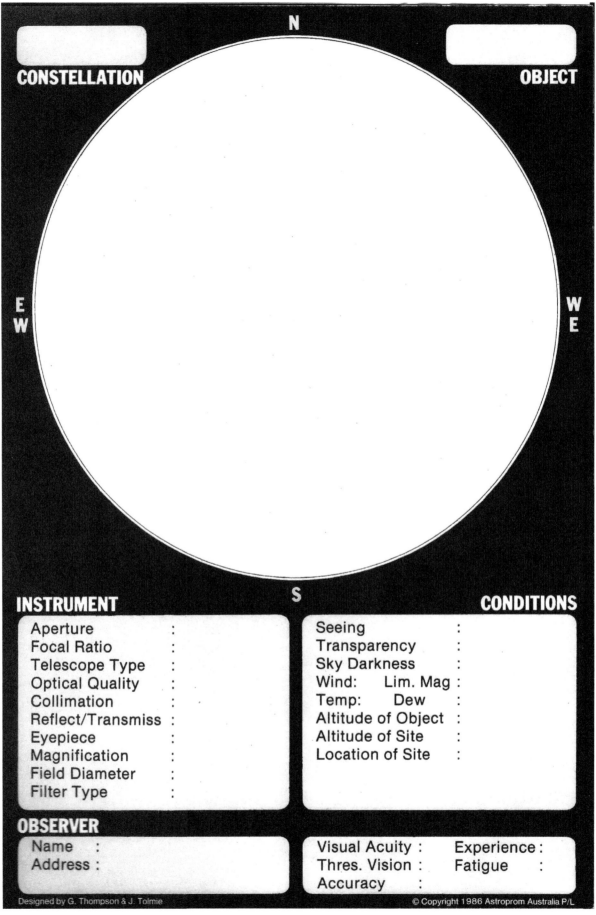

Chapter 11
PHOTOGRAPHING THE HEAVENS

Astrophotography has become a very popular hobby for many amateur astronomers. You will see superb examples of it displayed throughout STARGAZING. These may entice you to take your own astro-photos, if you haven't already. Be warned; once you get into astrophotography, it can become an addictive hobby!

In the mid-20th C, photography of the sky was very limited. Almost all of it was taken with black and white film. In those days, there were very few professional astronomers, and there were very few large professional telescopes. Professor Sumner Miller famously stated that "*The large professional telescopes were all in the northern hemisphere, and most of the best objects are in the southern hemisphere!*" And of course, there were no space telescopes back then. But nowadays, there is an abundance of giant telescopes in both hemispheres and numerous space telescopes and space probes.

In the 1960s, amateur astrophotography had barely started to emerge. There were very few reference photos for deep sky objects in the limited number of astronomy books that were published then. But today, advanced amateurs using the latest computer enhancement programs, are capturing superb, colorful, high-resolution images, the best if which can surpass many of those taken by the world's largest professional observatories! Due to the Internet, we now have access to numerous extraordinary images of space. It is a golden age for stargazers!

There is so much to say about how to go about telescopic astrophotography that many books have been written on it. Should you be interested in trying your hand at it, this chapter covers the basics to get you started, and it will give you some ideas about what types of astrophotography you can try.

Ace amateur astrophotographer Don Goldman summed up the art of astrophotography beautifully by saying: "*Our hobby is technically demanding to keep the telescope pointed on a star all night long and to keep the system in focus as the night cools. However, the end result is art. We attempt to frame our object and colorize it in an aesthetically pleasing manner to present the awe and mystery of the universe. Like famous artists using oil paints on canvas, astrophotographers all have different styles, neither right nor wrong, just expressive. Therefore, I firmly believe that what we do in this hobby is 'technical art'. We are landscape photographers on a cosmic scale.*"

It seems that most astrophotographers tend to be best at photographing a particular type of object, be it the Moon, the Sun, the planets, comets, the Milky Way, or particular types of deep sky objects. Very few are good at excelling in all types of astrophotography because each type requires special skills.

THE BASICS

For astrophotography, you'll need **a camera with manual settings**. Most brands have these. A Single Lens Reflex (SLR) camera is the best type to use. Because cameras use battery power, you might not be able to take too many timed exposures before the battery runs down, especially if the camera gets cold. Coldness will make the camera battery die but it will recover when it returns to room temperature. To be safe, have some spare, warm batteries on hand.

If you are doing star trail photography with long exposures, or even short ones, you'll need a **sturdy tripod** to hold the camera perfectly still so that vibrations from a breeze do not blur your image.

To start off, set your camera to the **manual setting**. This setting is great for star trail and lightning photography. Most standard 'point and shoot' digital cameras do not allow for exposure times over 15 to 30 seconds, so they will not be any good for long exposure photography, which requires from several minutes up to hours. So you don't risk bumping the camera out of position, or blurring the image when you press the camera shoot button, use a **shutter release cable on an SLR** to start the exposure.

To **focus** on the stars, turn off auto-focus and select infinity. You may need to vary the lens focus just a little to get the sharpest star images. Seldom is the best focus exactly at the infinity marking on the lens. The perfect focus may be slightly inside the infinity symbol so experiment by taking a number of images with a different focus slightly inside the infinity marker being sure to record which position the camera lens was at for each exposure. Select the sharpest image when they are enlarged, and then mark that position on the lens for future shots.

Because the majority of stars and the Milky Way are faint, you will need as much light as possible to record them, so the larger your **camera lens** is, the more faint stars and nebulas you will be able to record. A large diameter lens will also shorten the exposure time and it will make your images sharper.

The **aperture setting** on the lens is what controls the amount of light entering the camera. Because most stars are faint, you will need to let in as much light as possible, so use an aperture setting of f4 or f2.8. An aperture of f2.8 will let in more light than the f4 setting, but the stars will not be as sharp.

On your camera, use a high **ISO setting** (the equivalent of film speed) such as 400, 800, or 1600 to maximize your camera's sensitivity to faint light. There is a trade-off however; the higher the ISO settings, the more graininess there will be when the image is enlarged. Some cameras have an ISO setting as high as 3200 but this is too grainy for astrophotography.

The **exposure time** affects how much light the camera receives. For faint stars, nebulas, and the Milky Way, you will require exposure times of several minutes. But during that time, the Earth will have rotated, so this will cause the stars to leave trails when using a tripod. However, star trails can look very interesting. There are some star trail images in this chapter and elsewhere in STARGAZING. In order to not have stars leaving trails, you will need to mount your camera on an equatorial mount that has a motor drive that compensates for the Earth's rotation. If you have a motorized telescope, simply attach your camera to it. Digital cameras are much more sensitive to light than film, so exposure times are much shorter.

Another variable to consider is the **focal length of the camera lens.** If you want to photograph a wide area of sky to capture a large constellation, perhaps a large portion of the Milky Way, or a comet with a long tail, then you will want a low focal length such as 24 mm, or 18 mm to get a very wide field. If you want to photograph the Moon and have it show some detail, you will need some magnification. This will require a telephoto lens with a high focal length of at least 200 mm. An SLR camera will allow you to use interchangeable lenses that have different focal lengths, or you could use a zoom lens.

If the temperature is close to, or below freezing, then you will need a **lens heater** to stop the lens from dewing

Expert astrophotographer Daniel Verchatse seen here in his fold-back roof observatory in Chile.

over. Blowing warm air over the lens every so often using a hairdryer is a practical means of warming the lens. A long lens hood also helps to eliminate dew. Alternatively, place the camera under a shelter where it can still see the part of the sky that you wish to photograph.

When starting out, experiment with different exposure times, ISO, focus settings, and white balance to see what works best for different subjects.

If you are photographing from a suburban or city location, your photos of the sky will have a blue or brown background glow caused by city street lights. To avoid this, go to a dark country location where you will have a naturally dark sky.

There are numerous books available on advanced astrophotography and all are very good. **Alan Dyer** has written a 500 page, multi-touch ebook entitled *'How to Photograph and Process Nightscapes and Timelapses 2nd Edition'*. It covers all aspects of astrophotography in considerable detail and it has many examples of good astrophotography. For other books on astrophotography by other authors, see Appendix 3 on Outstanding Astrophotographers.

PHOTOGRAPHING STAR TRAILS

When you photograph the night sky with a stationary camera using exposure times for an hour to several

Above: **Star trail photography**

Due to the rotation of the Earth during this three-hour time exposure, stars in the far northern sky trail around the North Star at the far right. This magnificent picture was taken by Daniel Lopez with the El Teide volcano on Tenerife Is just after sunset using a very wide-angle lens. The color of stars is determined by their temperature. Red stars are relatively cool, yellow stars are hotter, and white stars are much hotter and blue stars are very hot while purplish-blue stars are the hottest of all.

Left: **In this 1 hour time exposure of star trails around the South Celestial Pole taken by Phil Hart, bioluminescence plankton can be seen in the waves lapping the lake's shore.**

The blue light in Macquarie Lake in New South Wales, Australia was caused by an exceptional blooming of plankton called Noctiluca Scintillans (see inset). It's 2 mm in size and it emits a blue chemo-luminescent light when disturbed by movement of the water caused by waves, boats, people swimming, and stones being thrown into it. The light is similar to that of fireflies and glowworms except that it is electric blue.

hours, you can record colorful star trails. Wide angle lenses give very impressive results.

Any exposure over 30 seconds will start to show a trail. The longer you leave the shutter open, the longer the trails will be. Point the camera in different directions to get different star trail arcs. The colors of the stars will be more pronounced than they are to the naked eye. Colors can be further enhanced using photo editing programs on a PC.

If photographing near a city in the early evening, you are likely to record trails from plane lights. Satellites and shooting stars may also cross your image.

For capturing star trails, you will need to take long exposures using the 'M' (manual) setting on a Single Lens Reflex camera (SLR). While the shutter is open, the camera will be recording light. With a digital camera, you may need to read the camera instructions to find out how to keep your camera lens open for long exposures.

To give your image perspective and to make it look artistic, include a silhouette of a tree or some interesting terrestrial object in the foreground and include the horizon to give it depth. You can light foreground objects, by giving them a short flash of light. Having terrestrial objects in the foreground can give the image a dramatic atmosphere.

In this marvelous image, we see the southern Milky Way from Vela on the horizon to the center of the Milky Way overhead as seen above four of the ALMA radio telescope dishes in Chile. The horizon and the dishes were added to the background image. Credit: ESO

CAPTURING THE MILKY WAY

Using digital cameras with a fast lens mounted on a sturdy tripod, spectacular shots of the Milky Way can be captured with relatively short exposures. When tracking the Milky Way, deep exposures can be taken over hours.

If you take an exposure over 15 seconds, then the stars will trail. To stop the stars trailing, you will need a tracking platform. But by tracking the stars, this will blur the landscape, as it did a little with the picture below of the VLA telescopes. To overcome this, you can take a long exposure tracked picture to capture of the Milky Way, and another short exposure picture of the foreground. This way both will be sharp. Using Photoshop, combine both images by overlaying the foreground image over the Milky Way image.

Left: **In this wide-angle view, an astrophotographer is standing on a rocky ridge capturing the center of the Milky Way on the horizon with the northern constellations extending to high overhead.** Credit: Mike Taylor

Right: **Here we see another view of the Milky Way extending from the center of our galaxy overhead to the red North American Nebula in Cygnus near the horizon. The bright star at top right is Jupiter. In the foreground is the Very Large Telescope in Chile. This image was taken by the highly accomplished astrophotographer, Yuri Beletsky.** Credit: ESO

Bjørnar G. Hansen in Norway captured this ethereal scene of a bright blue bolide streaking through the Big Dipper during a display of the Northern Lights. Mars is the bright orange star at the far right.

In the northern hemisphere, Iceland, Norway, Sweden, Greenland, Finland, Northern Russia, Alaska, and Canada are lands that offer magnificent auroral displays that can light up the whole sky, and the landscape as well.

PHOTOGRAPHING AURORAS

Photographs of auroras can produce some of the most dreamlike scenes one can capture on film. If you live at a latitude above 50°, you will get a chance to photograph some spectacular auroras on moonless nights when the Sun is active. When streams of atomic particles in the solar wind are drawn into the upper atmosphere along Earth's magnetic field lines, the energy these particles transfer to oxygen and nitrogen molecules in the air causes them to fluoresce to form auroras. For stunning images of auroras go to Volume 2 Chapter 1.

Auroras look you are standing under giant curtains hanging from the heavens that are very slowly moving back and forth from a slight breeze. Because they are moving, the will blur in long exposures, so you need to take as short an exposure as possible using a sturdy tripod that will not move in a breeze. Experiment to determine the best exposure times. Aperture settings around f4 will most likely give the best results. If you take photographs in very cold conditions away from civilization, be sure to have plenty of warm clothing, and be prepared for an unexpected snow storm by having a pre-planned escape route.

CAPTURING COMETS

Every few years, there tends to be a bright, naked eye comet that can be well photographed. You can get good images of bright comets using a fast lens mounted on a tripod, but you will achieve much better results by tracking the comet so you can take longer exposures. To do this, attach your camera to a telescope with a motor drive. In photographs, the comas of many comets display vivid turquoise and green colors. Their gas tail is typically blue, while its dust tail is a light creamy color. For many more images of comets, see Volume 2 Chapter 12.

PHOTOGRAPHING METEOR SHOWERS

To photograph meteor showers, you will need an expensive, fast, large aperture, wide angle lens on your SLR camera. This will permit you to capture a lot of light quickly. You need to do this because most meteors last for less than a second. The camera needs to be mounted on a motor drive that will track the stars in order that you can take long exposures of more than an hour or two. The camera's field of view should encompass the part of the sky from which the meteors radiate. Because meteor showers can cover a large area of the sky, and sometimes the whole sky during a big shower like the Leonids, you will need several cameras on the one tracking mount. Their fields of view should overlap by 30% so you can make a whole-sky montage using an editing program such as MaximDL. This will combine the images from several cameras to form one large mosaic. For great images of meteor showers and bolides go to Volume 2 Chapter 2.

Colin Legg took this stunning shot of Comet Lovejoy in December 2011 before dawn from outside Perth, Australia.

Here we see **Comet Neat** photographed on the nights of April 18th and 19th, 2004. The coma appears distinctly green coma. The image on the right shows a kink in the tail caused by the tail interacting with the solar wind. Credit: Loke Kun Tan (StarryScapes)

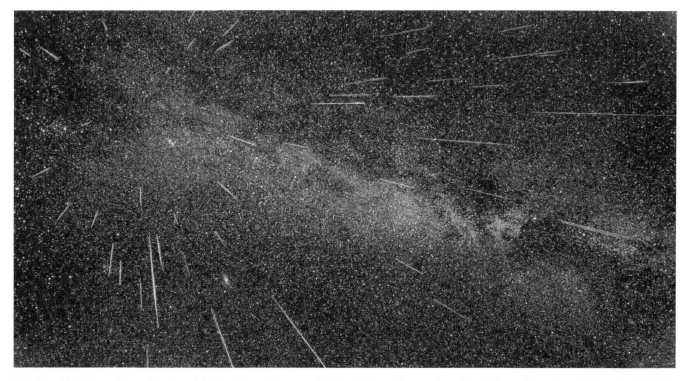

The Perseid Meteor shower is often rich in meteors. It can only be seen from the northern hemisphere due to it radiating from the far northern constellation of Perseus, which is circumpolar. Note that these meteors start out as green and then become magenta as they penetrate deeper into the atmosphere.

In this image, note the Perseus double star cluster lies just below the radiant at top left. The Andromeda galaxy, M31 is the small, bright fuzzy ellipse below the Milky Way to the left of center. Credit: Fred Bruenjes

CAPTURE THE MOON IN COLOR

It is fascinating to photograph the Full Moon in color using a good telephoto lens. To get high resolution, try photographing the Moon at the prime focus of your telescope. Color on the Moon is imperceptible to the naked eye, but in a photograph, it can be enhanced by using a photo-editing computer program. A color-enhanced image reveals different colors in the lunar lava seas. This is caused by there being different mineral compositions in the meteors that formed the basins, and in later lava flows that flooded into them. (See Volume 2 Chapter 5, page 119.) Computer processing also allows one to improve the color balance and the contrast and to brighten faint regions as well as darkening the background sky to greatly improve an image.

When color photos of an apparently colorless Moon are enhanced, different colors emerge due to different mineral compositions in its crust. Credit: Eddie Trimarchi

PHOTOGRAPHING THE FULL MOON RISING OR SETTING

If you plan ahead, you can get some impressive shots of the Full Moon rising or setting over a distant landscape. To do this, look up on the Internet, the exact times that the Moon will rise or set, as well as where this will occur along the horizon, so you are prepared well in advance. You'll need to use a good telephoto lens mounted on a tripod. To give the Moon scale, it will look best if the Moon is rising or setting behind an interesting natural landscape feature, or some manmade structure in the distance.

Here we see a dramatic image of the Full Moon rising behind the observatories of the Very Large Telescope in Chile.

The purplish gray shadow of the Earth in the atmosphere is just above the horizon. Above this is the pinkish 'Belt of Venus'. This impressive image was taken only 10 minutes after sunset. Gordon Gillett took it from 14 km away from the observatories using a 500 mm lens.

PHOTOGRAPHING LUNAR DETAIL

The Moon is very bright, so you can use very short exposures to photograph it through your telescope. The images will therefore be sharp. You will need a camera adaptor fitted to your eyepiece holder. To minimize poor seeing, photograph it when it is high in the sky. Experiment with different magnifications. After some experimentation, you are likely to get some great results. See Volume 2 [Chapter 5](#) for outstanding lunar photography by amateurs.

At the Moon's terminator, the Sun is low in the lunar sky making even the most minor variations in elevation stand out starkly. This produces striking detail. In this shot, we see the large crater Copernicus at top right. Note the craterlets around it which were formed by ejecta from the explosion Across the center, there are many low relief, small hills that appear pointed by their shadows but which are not. Credit: Steve Hill

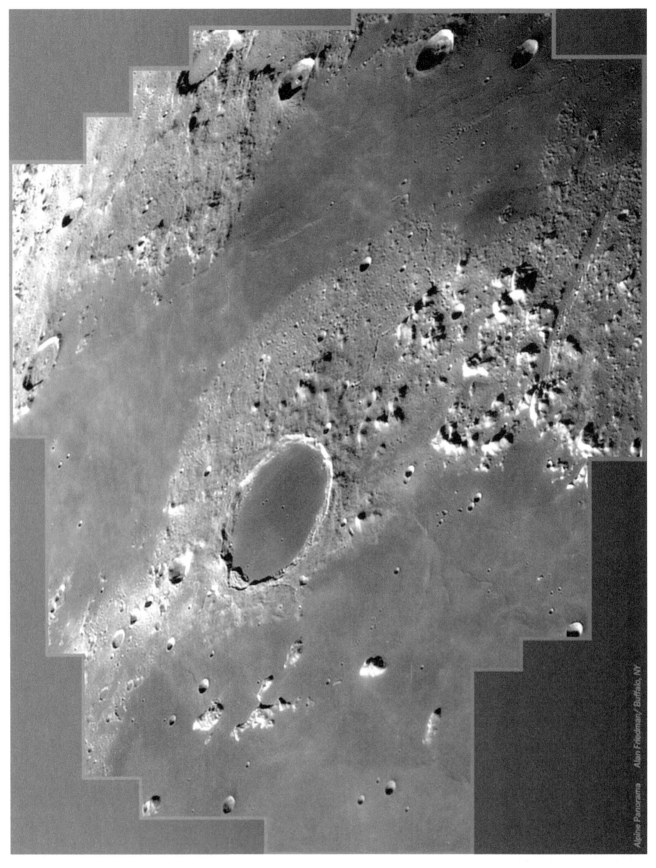

This spectacular, wide-angle view of the Moon by Alan Freidman in Buffalo, New York was achieved by combining many high-resolution images to produce this super-sharp montage. The image extends from the Alpine Valley (lower right) to north of the large, flooded crater Plato near center. In the original, on a large monitor, one can zoom right into Alan's pictures to see incredible detail.

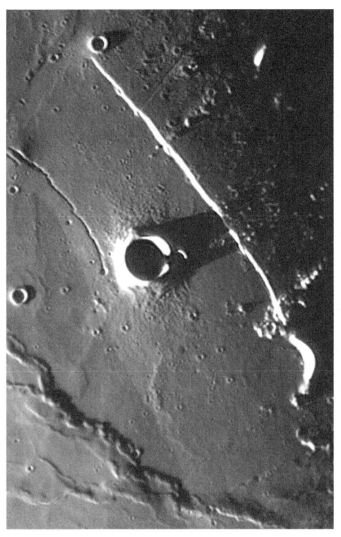

In this very high-resolution image, the white line is the lunar fault, Rupes Recta, commonly known as the Straight Wall. It was photographed by expert lunar photographer, Wes Higgins.

Taking photos of the Moon at a telescope's prime focus (with no eyepiece in place) can deliver impressive whole Moon images. Russel Croman in Texas, USA took this wide field photo of a ¾ Moon as it was passing by Mars (at top). Mars' icecap is visible on the planet's bottom edge. Pictures like this show just how large the Moon's apparent diameter is compared to that of Mars.

The rotation of Jupiter

Expert Australian planetary astrophotographer Anthony Wesley captured Jupiter's rotation and the movement of Io and its shadow in these superb, high-resolution images taken over an hour on July 25, 2007.

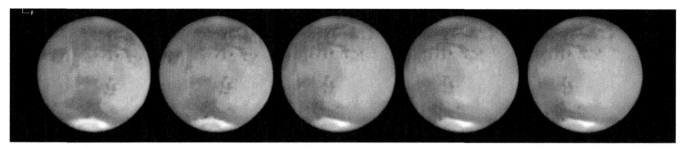

The rotation of Mars

Damian Peach in England photographed Mars several times over 3 hours during the night of Jan 26-27, 2010. Note how Mars is rotating to the left. These very high-resolution images show the dark region Acidalia adjoining the northern icecap. The gray water icecap can be distinguished from the very white carbon dioxide ice hood. There is a polar haze over the South Pole. If enough of these images can be captured, they can be turned into a video of Mars rotating.

PRODUCING HIGHLY DETAILED IMAGES OF THE MOON AND PLANETS

It's best to take up to hundreds of shots of the Moon and the planets one after another and delete any that are blurry due to poor seeing. A computer photo-editing program will do this automatically for you. Programs like MaximDL will stack the best shots together to greatly improve the clarity of your photos. This will increase the contrast and make detail much more obvious. If bad seeing has moved some image pixels out of alignment relative to those in other images, the program will move them into place. This goes a long way to eliminating blurring due to poor seeing thereby producing extraordinarily detailed images. You can take overlapping images of the Moon at medium to high power and then have the computer seamlessly stitch them together into a much larger, highly detailed image.

If you take many pictures of Mars, Jupiter, and Saturn over a few hours, you will be able to run them as a video to see them rotate of these worlds. An extremely good example of this is available on Richard Bosman's website, Astro Fotografie. In his Mars section, scroll down to 'Mars rotates 2005, Winjupos 5 opnames and press animation. Richard's lunar images display astounding detail.

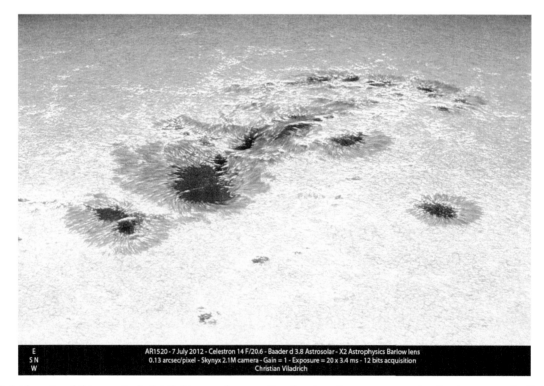

Christian Viladrich captured this amazingly detailed image of a magnetic solar storm known as sunspots when it was near the limb of the Sun.

RECORDING FEATURES ON THE SUN

Photography of the Sun can be exciting when you capture large prominences, unexpected brilliant white flares, and groups of sunspots evolving hour by hour and day by day. To photograph solar prominences, special solar telescopes with special filters are required. (For stunning solar imagery, see Chapter 7.)

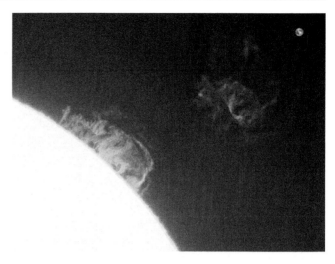

A solar prominence throws of gas into space. An image of Earth is shown for scale. Credit: Alan Freidman

DEEP SKY PHOTOGRAPHY

Deep sky photography is very popular because there are so many colorful objects to record in each category. Large nebulosities require a wide field, while small objects like distant planetary nebulas and galaxies need small fields obtained with fairly high magnification. The type of telescope or lens you have will determine what objects will be best for you to photograph. To get the most out of your images it is important to learn how to massage their faint regions and bright regions, and to get the color balance just right. It can take some years of experimentation to perfect your deep sky technique. There are many books available on deep sky photography to help you learn from the experts. Many amateurs have websites that display their work. Go to Appendix 3 to see many of the best astrophotographers in this field. Volume 3 displays numerous extraordinary examples of the best deep sky photography taken by many of the world's best astrophotographers.

This remarkable super wide-angle mosaic image was taken by Greg Bradley. It shows the large red arc of Barnard's Loop. At left of center are the two small blue reflection nebulas with an intervening dark nebulosity. This is M78. The Horsehead and the Flame nebulas are above center and M42 and M43 are in the upper right.

Rich Bowden of Bay Top Observatory in South Australia took this superb image of IC 2944 showing globules of dark nebulosity that may be in the process of forming solar systems.

Steve Cannistra captured this superb wide-angle image of M31.

NIGHTSCAPE PHOTOGRAPHY

It is a popular trend to take magnificent wide-angle images of the Milky Way with it acting as an atmospheric backdrop to an imaginative foreground. Many examples of this are displayed throughout *STARGAZING*. They require a large, fast lens.

Tony Dallas captured this extraordinary desert nightscape that shows a deep image of the Milky Way. Note the dark streamers converging on the Antares region.

Credit: Michael Goh

This marvelous image was taken by Xiaohua Zhao in Bolivia on the salt lake flat Salarde Uyuni. It has a very shallow covering of water to reflect the stars.

movies are shot by taking a long series of exposures automatically over a period of typically several hours. For example, you may take an exposure every 10 to 30 seconds over several hours and by using a simple computer program, it will convert your stills into a movie. In this way, you can record the constellations, the Milky Way, and occasionally naked eye comets moving across the sky as the Earth rotates. The Moon rising or setting looks great as a video. Clouds passing over and lightning in storm clouds add drama. Some astrophotographers place the camera on a traveling base that runs along a track so the camera can move across the landscape to give added interest. The camera can also be made to move slowly upward into the sky for more interest. Timelapse photography allows us to see things that we would otherwise never have noticed.

A supernova in the spiral arms of the beautiful galaxy NGC 134, was discovered by New Zealand amateur astronomer, Stu Parker. The supernova is marked at the top right in this color image. Credit: ESO's Very Large Telescope (VLT) at Paranal. Chile

THE NIGHT SKY AS SEEN WITH TIMELAPSE VIDEO PHOTOGRAPHY

More and more, amateur astrophotographers are now turning to time-lapse photography and running it as a video to achieve extraordinary results. Time-lapse

PHOTOGRAPHING GALAXIES TO DISCOVER SUPERNOVAS

Amateur astronomers searching for supernovas (exploding stars in other galaxies) take hundreds to thousands of black and white digital images of galaxies in fast succession each night using a computer program that automatically finds and photographs each galaxy. (See Volume 3 Chapter 2 & Volume 2 Chapter 12.)

Chapter 12
OBSERVATORY DESIGNS FOR AMATEURS

The large roll-off roof observatory of the Astronomical Society of Northern England houses a large refractor (gray) and large Newtonian reflector (white). Credit: Astronomical Society of Northern England

WHY BUILD AN OBSERVATORY?

Increasingly more amateur astronomers possess large telescopes that are impractical to carry outdoors and set up and then put away. In view of this, they need to build observatories in which to house their telescope. This chapter looks at the pros and cons of various design options.

The advantages of observatories are:

1. They provide **protection from the elements** - wind, rain, dew, and stray light. If a quick rain shower comes over while observing, there is no rushing to pack up. You simply close the roof.
2. They are **good security for your equipment.**
3. They **save a lot of time by not having to set up and pack away what is needed for an observing session. The time saved can be spent on observing.** An observatory makes it unnecessary to carry out and put away a telescope, a table, an observing chair or ladder, and items such as power leads, eyepieces, accessories, cameras, a laptop, star charts, reference material, and notepads etc.
4. **In an observatory, everything is at hand the moment you open the door,** so you are much more likely to want to observe or do astrophotography. To finish up, you merely close the roof and the door and switch off the power.
5. They eliminate the need to store telescopic equipment and all your accessories inside the house or in the garage.
6. **An observatory can incorporate storage shelving, a work desk, a bench with a sink, a small refrigerator for beverages, and bench seating for visitors – practical things that make it comfortable.** One friend has a media center in his. An observatory can also have soft carpet, so your feet don't hurt if you are standing for a long time. They can also incorporate lighting that is practical for observing, as well as a bed to lie on to read when waiting for your astrophotography exposure to end can be useful.
7. **They can be a great retreat during the day or on cloudy nights.** They can act as a study for carrying out research, or for working on your computer to massage your astrophotos. They are good for doing maintenance and other work undisturbed.
8. **They project an image of an organized, serious astronomer.** Observing in the open in one's backyard, on the footpath, or in a driveway doesn't have the same cachet as an observatory. And nor does it give you privacy like an observatory does.

THE SIZE OF AN OBSERVATORY

The size of your observatory may depend upon whether you need it for photographic or visual astronomy. If it is just to house a telescope for astrophotography, then you will probably only need a small space that is only a little larger than the telescope. If it is for visual astronomy, then you will want it to be large enough to move around with ease. Of course, your budget will determine the size that you can afford.

Some of the most memorable times I have had observing have been when I was sharing a marvelous view in the eyepiece with other astronomers, family, or visiting groups. If you would like to have others join you from time to time when observing, then you should have an observatory that is larger than a small garden shed. Extra room will pay big dividends over the years. You will need enough space around the telescope to make it comfortable for you, the telescope, and your visitors to move around easily. It's good to have some fold-up chairs or benches for people to sit on while they are waiting their turn to look through the telescope. Visitors will enjoy the experience much more when they are comfortable.

Plenty of room is especially necessary for a club observatory. You may decide to provide public observing nights to raise funds for more equipment, so plan ahead to have enough space to do so. It is not wise to cut your building costs to the bone because, in most cases, it does not cost a great deal extra to have more space. And keep in mind that it's often the case that you will have more money available in coming years.

Small garden shed roll-off roof observatories are the most common. Because space in these is limited, it makes it difficult to have much more than one or two other people in the observatory with you, and there is little opportunity for them to be comfortable. However, if the observatory is only going to be used for astrophotography and not visual observing, then a small observatory will suffice.

DESIGN OPTIONS

The three main designs for observatories for most amateurs are:

1. a **dome**,
2. a **roll-away roof garden shed** and,
3. a **dual roll-off roof observatory** - the most practical by far but this design is relatively unknown,
4. **Dual roll away sheds** – seldom ever considered as a possibility.

During most of the 20th century until the late 1980s, many amateur astronomers who wanted an observatory, typically built one with a dome - if they could afford it. Until the 90s domes were often chosen by those with large amateur telescopes because they looked like a small version of the Mt Palomar Observatory - the quintessential image of an observatory in those days. The Mt Palomar Observatory housed the world's largest telescope, the 200" (5.08m) Hale reflector which became operational in 1948. It remained the largest telescope in the world until the Heck 1 telescope was built in 1993. (The Soviet Union built a 238" (6 m) telescope in 1976 but it was said to have performed poorly.) The 200" was featured in many science fiction movies. For most amateurs, this type of domed observatory encapsulated the essence of astronomy. (See Chapter 13, page 248.) But domes are not that practical for amateurs for many reasons, as we will see.

In recent decades, prefabricated garden shed observatories with a roll-off roof have become popular because they were fairly easy to construct, and they are inexpensive. But there are a number of disadvantages with this design that you may wish to factor in when considering a design for your observatory. After we look at domes and roll-off roof observatories in some detail, I will suggest some design options for a *dual*, roll-off roof design that has many advantages over either of the aforementioned designs.

TRADITIONAL DOMES

Professional observatory domes were originally designed to house very long, large refracting telescopes; but amateurs do not have such long telescopes, so the very considerable expense of a dome is unjustified to house today's relatively short telescopes.

Some well-constructed domes can look aesthetically appealing, but they have the following practical drawbacks that need to be considered when choosing an observatory design.

Domes have the following disadvantages:

1. **For most people, they are difficult to construct if you are planning to build your own observatory.** They have to be perfectly round in order that they do not bind on their rollers when they are rotated. This is a common problem for homemade domes. If they are made of metal, they often leak along one or more of their many seams due to summer heat expanding the metal and breaking the seals. In a storm, rain can come in through the shutters, or be driven up under the bottom of the dome by wind. This can cause costly damage to electronics and the telescope's optics.

2. **Domes are expensive to build, or to buy.** Commercially-made domes are expensive to buy and costly to transport. The larger a dome is, the more the cost skyrockets, and the greater the wind loading is in a storm. For the average amateur, they are too complicated to construct themselves. A dome is typically built on the ground and when completed, a costly crane is usually required to lift it into position. Prefabricated domes are usually well made but they are small and expensive for their size.

3. **Most domes have relatively small diameters to keep the cost down, so there is little space inside them.** Most domes provide very little free floor space because the area is typically only a little larger than the turning circle of the telescope. In most domes for amateurs, there is little space for anything but the telescope. Under the perimeter of the dome, head height is low. It is usually hard to have more than one or two people with you inside most domes.

4. **Many are noisy** when they are rotated at night, thereby causing sleeping family and neighbors to complain about the noise.

5. **They are typically not insulated, so they heat up a lot during the day, and in winter, they are very cold at night.**

6. **Dome can cause very annoying whirlwinds** inside them when the slit is open and facing within 30° to 60° on either side of the oncoming wind direction.

7. Domes create serious thermal problems that cause poor seeing. Heat from equipment as well as body heat rises through the slit in front of the telescope, thereby making the object being observed look blurry. With even one person under a dome, body heat can be a problem.

8. **A dome has to be frequently rotated** to keep the slit positioned in front of the telescope as it moves westward to compensate for the Earth's rotation. If the dome is not constantly moved, the telescope or camera will quickly end up looking at the inside of the dome! This often spoils photographs. A dome's narrow slit also limits the area of sky being photographed so wide-field photography is not possible. If the dome is not kept in front of the telescope, then the telescopic view dims and then disappears for the observer as the telescope moves towards the dome. This can easily happen when showing guests objects, but the owner does not know this has occurred while he is describing what the observer should be seeing. Guests say they can't see anything!

9. Domes have a slit that **only allows a small area of sky to be visible**. This makes it hard to find one's way around the sky if you are not using a computer-aided telescope. A narrow view of the sky takes away the beauty of seeing the night sky. The expanse of the Milky Way, the constellations, the planets, and meteors are not seen because they are seldom in front of the slit.

Dome observatories

Left: When there are treasured trees on your land or your neighbor's property that block your view of the sky, the answer can be to raise your observatory above the obstruction - as long as you don't mind heights! In such a case, a large water tank stand with a concrete floor worked well as a very sturdy platform for this observatory - even in moderate wind. In this observatory, a study/workshop is located in the room at ground level. A ladder is the means of entry into this observatory, which was built by Wally Best in Brisbane, Australia.

Right: Here we see Ed Ribson's small, domed observatory at ground level in Pittsford, NY photographed after a winter snow fall. Credit: Eve Strella

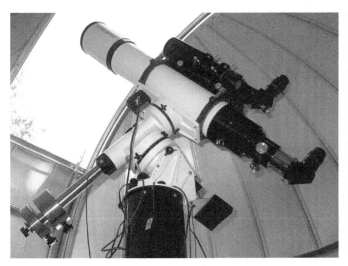
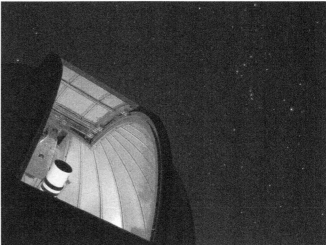

Left: **In a dome, there is a very limited view of the sky. Body heat and warm air from equipment has to rise through the slit in front of the telescope, thereby deteriorating the seeing.** Credit: Eve Strella

Right: **Dim red light is used in observatories to help the observer maintain some level of dark adaption.** Credit: Eve Strella

10. **The dome has to be moved each time the telescope is moved to another object.** This can be an annoying and time-consuming, reducing precious observing time. It can also be a strenuous job to move a large heavy dome manually, so domes often require the additional cost of being motorized.

11. **Domes can attract unwanted attention from inquisitive vandals.**

I have used many domes and found that for the reasons listed, they can be frustrating to use, particularly after I became accustomed to a much more practical dual roll-off roof observatory. Most modern, large professional observatories today no longer have traditional domes.

THE SKYSHED POD OBSERVATORY

SkyShed Pod is a commercially built, very small, prefabricated observatory. It features two half domes. The inner half dome is rotated inside the outer one. This offers an advantage over a traditional dome because it permits the observer to see half of the sky. When you want to look at the other half, you rotate both domes. However, having half of the sky open exposes the observer and the telescope to the elements and to stray light, particularly as the walls are very low. If it's windy, the observer can elect to open only the dome shells enough to see the section of sky required, but this has to be in the opposite direction to the airflow for the dome section to be of any benefit in reducing wind. With the half domes almost closed, this makes the slit very narrow towards the top. The design does not allow you to view the zenith at any time because both half-domes cover it. For automated telescopes that might be searching for supernovas, asteroids, or comets across most of the sky, the observer will need to be there to move the domes out of the way as the telescope moves from one quadrant of the sky to another. With all these limitations, a Sky Shed is little more than an expensive cover housing for a short tube telescope when it is not in use.

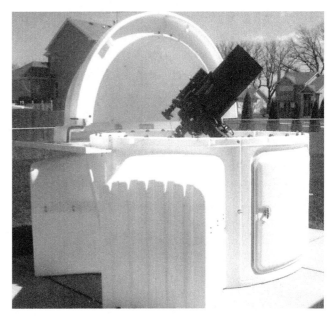

A SkyShed prefabricated observatory

The low height of the small diameter dome creates serious height restrictions. The door height is very low, so this requires the observer to bend down low to get inside. You need to be agile and be prepared to hit your head or scrape your back. Due to the low walls, it would work better if the floor inside the observatory was lower, but water-proofing the observatory to stop a sunken floor filling with water after a storm would be a challenge.

Because of the small amount of space inside the 2.4m (8') dome, it is only practical for owners who want to install a compact fork mounted Schmidt Cassegrain telescope. The Sky Shed is too small to accommodate a large Dobsonian, or equatorially mounted Newtonian telescopes. Note that in the photograph above, even short Cassegrain telescope sit above the SkyShed's very low walls, so the telescope and the observer are exposed to wind, dew, and stray light.

The SkyShed Pod costs considerably more than a larger, homebuilt, roll-off roof garden shed observatory that provides much more space and practicality.

Users comment that because the wall sections are not all the same height, the domes do not always rotate smoothly. The dome requires a lot of sealant to ensure it does not leak. It comes without a floor, so it has to be set up on a concrete pad, or a timber deck and bolted down so it does not blow away during high winds. The cost of the floor is a significant additional cost. The floor needs to be a little higher in the center so that water does not blow in under the walls and pool inside.

The shed's limited space is only comfortable for two people to stand around the telescope. It is more suited to those who are interested in astrophotography because they typically work alone, and they do not need to stay in the observatory all the time.

SkyShed Pod comes in a number of colors. The dome can be a different color to the walls. White looks attractive and it will reflect daytime heat - but at night, the inside of the dome and the interior walls reflect skylight and surround waste light, so they need to be painted a dark color - if the observer needs to maintain his dark adaption.

There is the option of adding one or more bays. These can be used to hold a set of drawers or a very small desk for a computer. However, you will have to bend down and sit very low to make use of them. The bays are designed to fit inside one another for shipping so they are not all the same size. This makes them look strange when assembled. Most observers will find little practical use for the bays.

SkyShed can be purchased for a few thousand dollars. Being prefabricated, it appears to be less work than building an observatory from scratch, but the DVD of the assembly instructions shows its assembly to be a lengthy job, and more difficult than one might expect. For more information, look at SkyShed's videos and user comments on the Internet.

A garden shed with a roll-off roof would be far more practical. It would provide more space and cost far less.

ROLL-OFF ROOF OBSERVATORIES

The advantages of a roll-off roof observatory are:

Home built roll-off roof observatories

Left: A typical garden shed, roll-off roof observatory.

Right: A large, expensive, custom-built, roll-away roof observatory for a large telescope.

1. relatively **easy to build,**
2. **inexpensive**, especially if a prefabricated garden shed from a hardware outlet is acceptable,
3. **all that is needed if it is merely to house a telescope,** especially if it is remotely controlled,
4. **quick to open and close,**
5. **seldom attracts unwanted attention from vandals.**

The disadvantages of roll-off roof observatories are that:

1. **most are small with little room for others to move around inside them** if they are the garden shed variety, so they are not good for having friends over, or public viewing nights unless they are specifically built large enough for that purpose,

2. **most do not have windows** so there is no natural light during the day unless the door or roof is open, and this can cause unwanted wind inside,
3. **most are not insulated** so they are very hot during the day in summer,
4. **they are very cold at night in winter** because they are typically fully open to the night air, and if they have metal walls the cold penetrates very quickly,
5. **they offer no protection from dew,** and it will even forming on metal walls when it is closed,
6. **the walls are typically not above head height so stray light can easily enter,**
7. they **offer very little protection from wind** buffeting the telescope, and making the observer uncomfortable,
8. if the garden shed types do not have sealed corrugated joins in the walls, and roof sealed, **dust blows in over everything when its windy**,
9. most are built as a low-cost option, so they are **small having have little room** for a desk with a computer to work at, a workbench for maintenance, storage for sundry equipment, seating for visitors, or a small tea and coffee making benchtop with a mini frig under it,
10. **few have the option of having practical, dimmable, low brightness lighting for observing, as well as bright even lighting for maintenance,**
11. they have **little use for anything other than housing a telescope**, so they have **no resale value** to anyone other than another astronomer, or a gardener,
12. esthetically, **they seldom compliment a nice home,**
13. **purely utilitarian providing no comfort** unless they are built large, insulated, and have useful head height.

Prefabricated, garden shed roll-off roof observatories are common. A three meter (10') square garden shed costs only a few hundred dollars. A garden shed is fairly easy to construct and it's sturdy enough to house a telescope. The roof section has to sit on rollers that run along two tracks. The rollers and track are purchased separately. The track typically extends the length of the roof, or a little more beyond the shed. This allows it to roll out of the way for observing the sky. The track is supported by two piers located near the end. The support columns for the track must be able to handle the weight of the roof when it is rolled away. It's best to place a brace between the track piers, so they always stay the right distance apart. The roof must be tested to be waterproof before lifting it onto the track with a few strong friends. The track design must ensure it cannot blow off in strong wind gusts. The roof should be locked in position when

A garden shed roll-off roof observatory used for astrophotography. Credit: Greg Bock

not in use. There are websites that give detailed directions on how to construct this type of observatory.

Astrophotographers often choose a roll-off roof observatory because they need little room other than that needed for their telescope and because it offers a whole sky view. Today, photographic telescopes can be controlled remotely by a computer so there is no need to make an observer comfortable inside an observatory. Via a computer, the astronomer can instruct the telescope to automatically open and close the roof if it is motorized, to acquire a predetermined series of objects and then photograph them. Weather detection sensors can be installed to automatically monitor cloud, rain, or high wind, and automatically close the observatory if necessary.

A simple roll-off roof observatory Credit: Ben Gaza

THE TELESCOPE MOUNT

A pier mount for the telescope is better for an observatory than a tripod because it provides maximum floor space around the telescope. The telescope should be bolted to a pier set into a concrete column dug at least a meter into the ground. Ensure that it is vertical and that the top is level. The floor can be timber decking,

Amateur astronomer, Peter Marples has built a climate-controlled room that is at one end of his observatory. While his telescope is photographing galaxies searching for supernovas, or while he is waiting for cloud to clear, he can watch movies, documentaries, television, search websites, send emails, or listen to music – all in comfort, regardless of the conditions outside. Credit: P Marples

Left: The plans shown here for the Meadowview single roll-off roof observatory allow for two telescopes and a workbench.

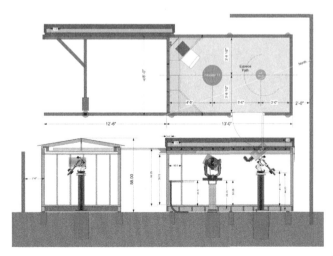

ply, or concrete. A timber or ply floor is less tiring on the feet than concrete. Placing secondhand carpet over either type of flooring will make the floor much more comfortable and this will reduce fatigue considerably. Underground power can be run to the observatory wall and the pier before the floor goes in. It's also useful to have a solid pathway to your observatory to keep the inside clean.

THE ROLL-AWAY SHED

Roll-away telescope sheds.

Left: *This simple roll-away shed parts in the middle and rolls along tracks to expose the telescope for observing. Credit: Gregg Thompson*

Right: *Jim Riffle's 18" fork mounted telescope was located at a remote site in Australia to photograph the southern skies. The rollaway storage shed rolls away on tracks. The telescope deck is at the same level as the house. Credit: Jim Riffle*

The roll-away telescope shed is not used as an observatory: it is a storage shelter for a telescope when it is not in use.

THE ADVANTAGES OF A *DUAL* ROLL-OFF ROOF OBSERVATORY

When I started to design my observatory, I considered the problems associated with domes and roll-off roof observatories. I set about seeing if I could design an observatory that could overcome each and every problem that other types of observatories had. I came up with a simple and practical design solution that I called a *Dual* Roll-Off Roof Observatory. This design has a typical roll-off roof broken into *two roofs*. This simple modification provides many benefits.

If built so that the roofs roll north south, and the walls are made to be above head height, these simple changes provide all of the following major advantages.

1. can **operate in full-sky open mode, or in slit mode,** which can be as narrow or wide as required,

2. makes an enormous difference to the **observer's comfort by reducing wind, cold, and dew** when the roofs are not fully open or in wide slit mode,

3. **eliminates stray light,** if the walls are the correct height,

4. allows **the telescope to track through a slit for much longer than a dome** without the roofs needing to be moved - as long as the roofs roll to the north and south,

5. **permits observations to be made low in both the north and the south** by rolling both roofs to one end or the other,

6. offers the **opportunity to have normal door height,** and to add in windows if one has a view, or one wants to daylight to work in the observatory without opening the roof,

7. **design can be large enough** to accommodate a number of people, as well as a desk, benches, storage cupboards and seating for others without the cost being high, and

8. **has the ability to become another useful room of a home that has aesthetic appeal and good resale value.**

The disadvantages are:

1. **it will cost a little more** by having two tracks with 4 posts. If the roof overhang extends beyond the end walls for a meter or more and is lined to stop wind coming in between the roof and the wall when the roof is rolled back, this will add some extra cost.

With this design, the observatory can work as both a slit-roof observatory for cold, dewy, or windy weather, and also as an open sky observatory when the weather is calm and warm. When it is a cold or windy night, the roofs only need to move apart a little more than the aperture of the telescope, however, on a moderate night, the slit can be as wide as required. On calm summer nights, both roofs can be rolled right back to provide a view of the whole sky. This is especially useful when you are moving around the sky a lot, or taking wide angle photographs. With the roofs fully open, the observer gets to see what is happening in the whole sky. Even a simple, inexpensive adaption of a garden shed to have two roof sections instead of one, offers many advantages.

This view shows a dual roll-off roof observatory with the slit open just a little wider than the aperture of the telescope. This provides the observer and the telescope protection from the elements.

Access to the observatory is via a staircase from ground level to the entry door, which is open on the side wall on the right. The door at the front center opens onto a roof deck that provides an excellent panoramic view of scenic mountains. Blackout blinds are pulled down over the windows at night to eliminate stray light. The blinds are up in this photo to enable the inside of the observatory to be seen. By opening the doorway to the roof deck and sliding the windows back, the telescope can view the western horizon. One roof rolls off to the north and the other to the south. Credit: Gregg Thompson

Here we see both roofs almost fully apart for open sky observing. Credit: Gregg Thompson

This view was taken during construction. It shows the tracks that the northern roof moves along.

Creepers later grew over the upright supports. When the roof rollers are placed a third of the way along the length of the roof at each end, the tracks do not need to extend the full length of the roof for it to be fully open. A third of the roof overhangs the end of the tracks when it is fully extended. A large workshop and storeroom were built in under the observatory. A 300mm diameter concrete pier extends from 1.5m below the workshop's concrete floor up to the floor of the observatory. The pier is independent of both floors. Credit: Gregg Thompson

THE VERSATILITY OF A DUAL ROLL-OFF ROOF OBSERVATORY

Closed

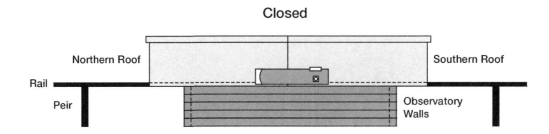

Slit Mode

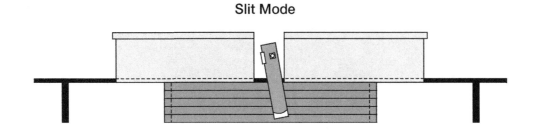

Open Sky Mode

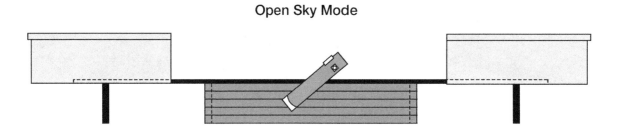

Observing low above the Northern and Southern horizons

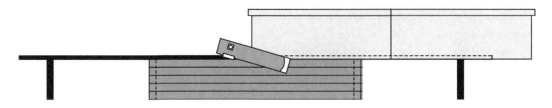

Observing low above the Eastern and Western horizons

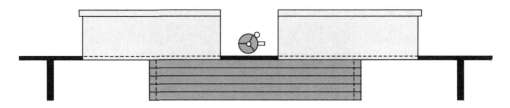

OPERATING IN SLIT MODE

When using the dual roof in slit mode, both roofs can be moved to place the slit where the telescope is pointing. The lower the object is in the sky, the wider the slit will be. It is best to have a generous overhang on the north and south ends of the roofs so they can move well inside the end walls of the observatory without allowing a breeze to enter the observatory through a gap between the end of the overhang of the roof and the north or south wall. This requires the underside of the overhang to have a ceiling.

SEEING THE NORTHERN OR SOUTHERN HORIZON

Both roofs can be moved to the same end of the observatory if there is a need to view an object quite low in the northern or southern sky. If you prefer high walls to keep out stray light, you can view close to the southern and northern horizon by installing barn doors in the end walls of the roof sections. However, it is uncommon to need to observe close to the horizon due to the atmosphere there being dense and often somewhat polluted, thereby making objects low in the sky appear dull and blurred.

WALL & ROOF CONSTRUCTION

If there is a possibility of heavy snowfalls, then the roofs need to be peaked. If heavy snow is not a problem then the roofs can be flat with a shallow, 15° slope. A flat roof overcomes having the peak of the roofs blocking out a part of the low northern or southern sky. Flat roofs are better in a windstorm because they have a low wind-loading factor. The low side should face the prevailing wind so that the wind will help force the roof down. The roof must be constructed so that the roof cannot lift off the tracks in a wind storm. (See drawing page 213.)

WIND PROOFING THE INTERIOR

If your observatory is located where it can suffer windy conditions, then it can be practical to install sliding door frames on the sides of the roof sections to stop crosswinds. If wind and cold can be troublesome, then install sliding panels to the top of the roof as per the sketch following.

INTERIOR DESIGN FINISHES AND LIGHTING

An observatory can be made to look very attractive with a little thought, and at only a little additional expense. Lining the interior walls and ceiling costs more, but it creates a homely feel, and it reduces the penetration of cold and dew.

It is very practical to have dark, matt interior finishes for visual observing so as to minimize the reflection of stray light and sky glow in order to maintain one's dark adaption. By placing some inexpensive astromurals, or space posters on the walls, this beautifies the interior

If your site suffers regularly from cold wind, then install sliding shutters on the sides of the east and west walls of the roof will largely overcome this problem. If cold breezes come in through the roof, they can make observing unpleasant, so attach sliding shutters on the top of the roofs as well. The roof shutters remain clipped together on the underside when the roof is moved apart. They can be uncoupled at any time and slid back. Credit: Gregg Thompson

with an astronomical feel. I splattered spots of white paint onto a satin black paint finish so they appeared like stars.

White light is required for maintenance and when working at your desk, so install dimmable, white LED strips at opposite sides of the observatory. It is best to have a palmet cover over these to direct the light downward, so it does not directly enter the observer's eyes. The palmet should swivel so that direct light can illuminate the rest of the observatory when working

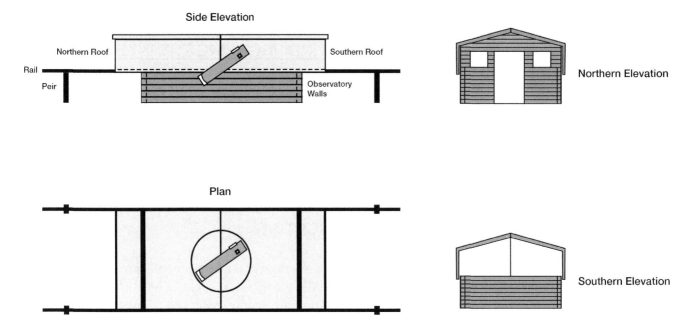

Plan and elevation views of a simple Dual Roll-off Roof Observatory

The observatory should be constructed so that the slit runs east/west. The advantage of this is that there is then little need to move the roofs to track objects when the observatory is operating in slit mode because the stars move more or less from east to west unless you are observing close to the celestial poles.

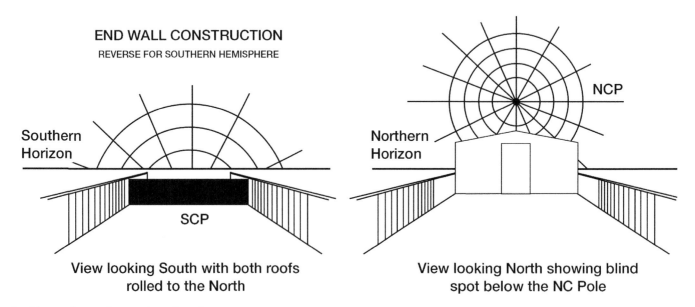

Looking at the northern and southern horizon

When both roofs are rolled to one end, a clear view of the opposite horizon is achieved. The north and south walls should face due north and due south if possible. One wall will be directly under your celestial pole. Stars will rotate around your celestial pole, so those in the east will move higher into the sky as the evening progresses making them visible over the wall in that direction. This wall can therefore be a traditional height in which a normal-sized entry door can be installed if you have side walls that are not high enough for this.

on repairs or adjustments to the telescope. Dimmable red lighting is also necessary to maintain the observer's night vision when making observations. It also helps to create an atmosphere when guests visit.

Windows are useful to let in daylight without the need for artificial light or to have the roof open. They are a benefit if you have a good view. Windows may require block-out blinds or curtains to stop stray light from entering the observatory at night.

1.

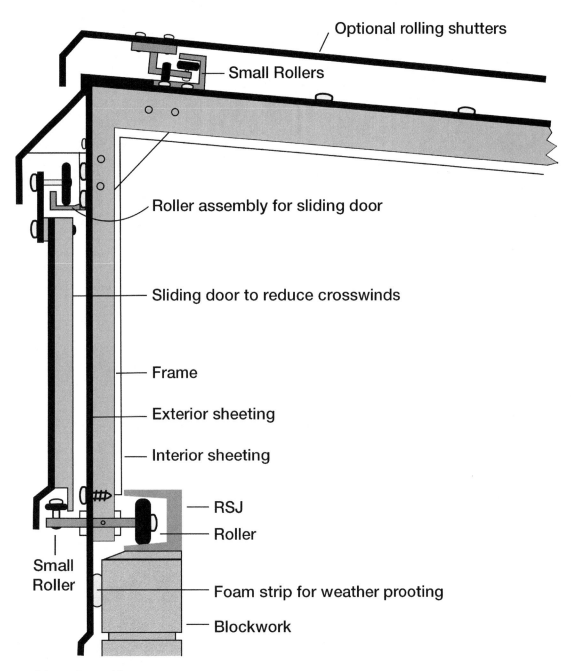

A cross-section of the roof assembly

This diagram illustrates a design for securing the roof rollers to the tracks, which are fixed to the top of the walls. The roof rolls on rubber wheels inside a C section of a steel beam commonly called Rolled Steel Joist (RSJ). The slight angle on the horizontal sections of the RSJ makes the wheels track perfectly, as long as the tracks are parallel. The angle also ensures that rainwater does not pool in the track. Rubber wheels make the operation quiet, as long as the roof is of a sturdy construction so that it does not rattle. The roof cannot lift off during a wind storm as the RSJs are bolted securely into concrete filled, steel-reinforced walls, or alternatively, timber stud walls. The sides of the roof walls should hang well below the track to stop rain penetration from storm winds. Rubber or foam seals will stop wind, rain, and insects entering the observatory when it is closed. In this drawing, an optional sliding door frame can be fitted to the sides of the roof to eliminate breezes passing through the slit opening section.

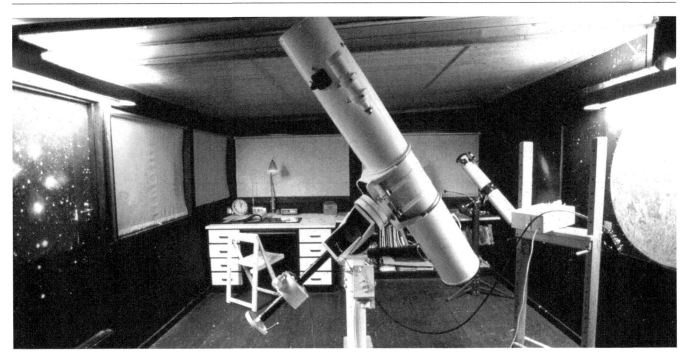

In my youth, I built this **Dual Roll-off Roof Observatory**. This is an interior view when it was first completed. **It housed my 200mm (8") telescope, but in later years it was home to a 350 mm (12.5"), and after that, a 460mm (18") telescope. Second-hand black carpet was installed, which made it much warmer in winter. It also made it much more comfortable on one's feet when standing at the telescope for hours on end observing. Behind the camera is a divan lounge for seating and resting.** Credit: Gregg Thompson

In this view, both roofs are partially opened to allow the telescope and the observer to gaze skyward. By not having the roofs fully open, this eliminates stray light and it minimizes sky glow when observing faint objects. To minimize cold in winter, it is best to have the slit only a little wider than the diameter of the telescope tube. A narrow slit also eliminates dew inside the observatory. Insulated walls and ceilings reduce cold substantially. Credit: Gregg Thompson

DESIGN VARIATIONS FOR QUALITY CONSTRUCTIONS

If a quality roll-off roof observatory is well designed for a good property, then it will add value when one decides to sell up. It can be sold as a playroom, an entertainment area, a granny flat, a study, a pool room, an extra bedroom, or a sky lounge. As a bedroom option, one can lie in bed and look at the stars or sunbathe during the day in privacy. By comparison, a domed observatory or roll-off roof garden shed, will add little additional value to the property – unless, in the unlikely probability that a potential buyer also happens to be a stargazer who wants an observatory.

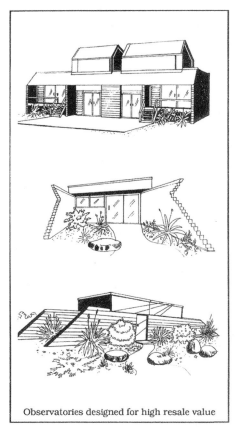

Observatories designed for high resale value

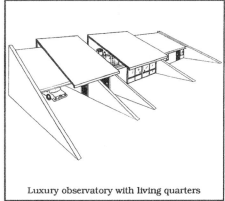

Luxury observatory with living quarters

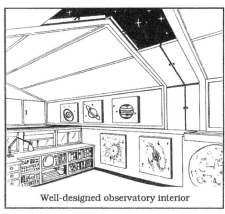

Well-designed observatory interior

Astronomy Club House Observatories

The dual roll-off roof design is a practical solution for astronomy club observatories that are used by multiple members or for public viewing nights because they can hold a number of observers if they are not too small. And they allow body heat to escape easily.

The up-market concept plans following are for an astronomy club observatory that has many active members who have gained substantial funds, possibly from a wealthy individual's estate, local businesses, and/or from the district council, in order to build the observatory. A progressive astronomy association could gain some government contribution on the basis that such a facility would benefit the local community and schools. Large donations are possible when the word is put around about what the astronomy club plans to build. Some wealthy donors would love to have their name live on in the stars by having the facility, or parts of it named in their honor. The design could include a large observatory on top with multiple instruments plus store rooms and a workshop under the roll-off roof sections. At ground level there would be a meeting room/ theatre, with a kitchen/cafeteria, an office, and amenities.

Dual roof observatories can have many practical design variations when a quality building is required.

The top left drawing shows an option for a country retreat with an observatory on top. The living quarters are at ground level in the center and the bedrooms are up half a flight of stairs at either end. Another half flight of stairs takes one to the observatory. The drawing at top right shows how an observatory for a large telescope can be built with a bedroom at one end and a garage or living room at the opposite end. The rolling roof sections of the observatory roll over the rooms at either end. The sketches below on the left show two simple modern designs for a roll-off roof observatory in a garden setting. The interior drawing at lower right shows what a large observatory may look like if it had sliding wind shutters installed on both the roof and the walls of the roof sections. Doors at the end of the roof can open to look low in the sky at either end.

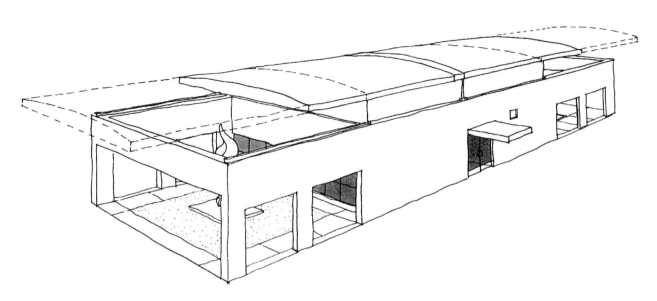

A concept sketch of a two-level clubhouse that incorporates a Dual Roll-off Roof Observatory on the top level to house two large telescopes and a binocular chair.

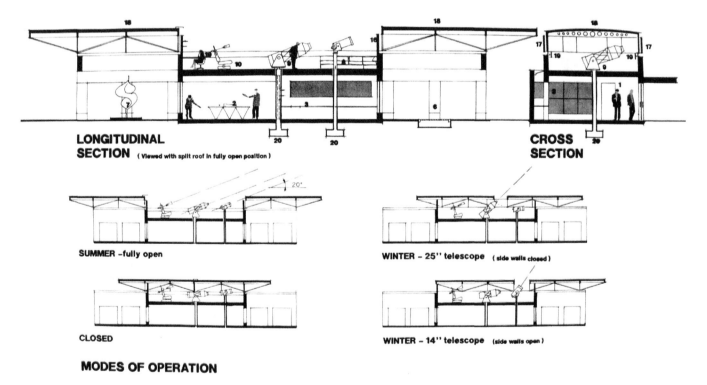

These drawing show indicative plans for a large multi-functional observatory for the use of club members, and for public viewing nights. The split roll-off roofs provide open sky viewing as well as slit operation. Credit: Gregg Thompson, R Biscoe

A COMMERCIAL DOUBLE OBSERVATORY FOR PUBLIC VIEWING NIGHTS

Some enterprising stargazers may want to turn their passion for showing people around the universe into a business venture by operating a public viewing observatory. To accommodate enough people on weekends and holidays periods, it could be functional to have two observatories side by side with a ticket sales counter and merchandising shop between them. Queuing lanes would be on either side leading into the observatories. During times where there may be fewer numbers during the week, only one observatory need be opened. Dual roll-off roofs are very practical for such use, as they offer the option for open sky viewing. This minimizes poor seeing by allowing body heat from patrons to escape easily. Body heat is mostly visible when viewing the planets and the Moon at high magnifications. Deep sky objects are typically viewed at low to medium power, so the effects of poor seeing are limited. On cold nights, the observatory could operate in slit mode.

A pair of dual roof observatories designed for a themed, edutainment stargazing facility for the public.

Both observatories have curved roll-apart roofs. They feature one or two large telescopes in each. The exterior could be painted in fluoro paint and lit with UV light, so it appears to glow at night. This soft lighting would be very attractive, and it would allow patron's eyes to dark-adapt to the night sky. Fiber optic lighting throughout the garden and along the pathway would provide a spacey entry statement, especially if accompanied by space music playing in the garden.

The interior of a dual roof commercial public observatory could accommodate one or two large telescopes. A large projection screen at one or both ends end could project videos about the objects being viewed in the telescope. Fold-away seating would permit those waiting their turn to look through the telescope to sit comfortably while watching video or stills of various objects under observation. In overcast conditions, the screen could feature a space documentary video. Low brightness, special effects lighting in the walls along the sides could incorporate soft, colorful images of space objects floating around to create a futuristic ambiance. When viewing faint deep sky objects, almost all lights would be off or dimmed to a minimum with no direct light being visible.

THE SPLIT SHED ROLL-APART OBSERVATORY

A Split Shed Roll-apart Observatory is very practical for housing large instruments. It is a low-cost solution to house one or more large telescopes. It would be an excellent option for astronomy clubs that have a dark sky site.

This design is a variation on the Dual Roll-off Roof Observatory. In this option, the entire building is placed on rails, so it can roll apart with one half going to the north and the other to the south. It has the option of rolling both halves well away from one another for open sky observing when required. On cold, breezy nights, it would operate in slit mode with the sections only slightly opened. To eliminate crosswinds, the side walls can have sliding panels added to the exterior, as in the top drawing opposite.

This design allows the telescope to observe objects low in the eastern and western skies. For observing objects low in the northern or southern skies, the building can have doors in the end walls. Both sections can roll to either the north or the south for as far as the tracks will permit.

With low friction rollers, the buildings could easily be moved by one person. A large concrete platform could extend well beyond the building to accommodate other non-permanent telescopes.

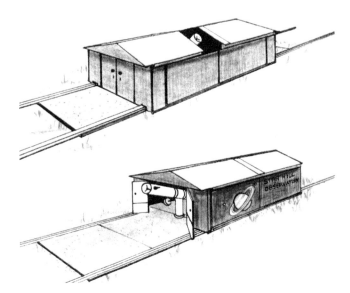

This roll-away split shed design offers many advantages for a club observatory in a county location. It provides a large volume of space, full sky coverage right down to the horizon in every direction as well as protection from wind and stray light. It is a relatively low-cost construction that provides security and great versatility. Credit: Gregg Thompson

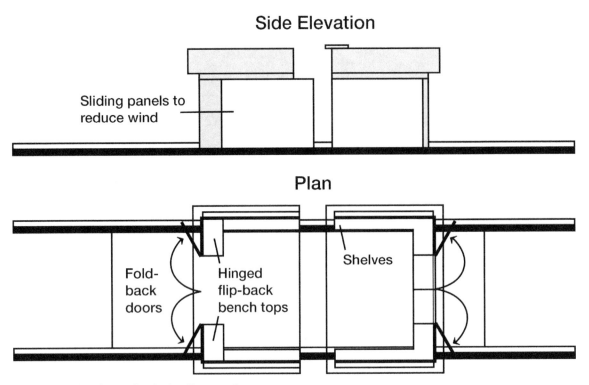

Plan and elevation sketches for a Split Shed Roll-Apart observatory

COMMERCIAL ATTRACTION CONCEPTS UTILIZING DUAL ROLL-OFF ROOFS

For those who may have an interest in developing a commercial venture to take people on a naked eye journey across the night sky, I designed a simply-constructed, roll-off roof 'sky theatre' that makes this a very special experience. In this attraction, which I named the 'Space Lounge', patrons enter into an indoor roll apart dual roof building where they find themselves in subdued lighting surrounded by a night scene at the end of twilight with stars in the sky above them. This is achieved by a mural artist painting a realistic night horizon scene around the walls that are curved in the corners. A realistic night sky is painted on the curved ceiling so that it looks like the stars overhead. The murals on the walls and ceiling are painted in UV glow-in-the-dark paint and lit by UV tubes hidden behind pelmets at head height so that visitors cannot see the light source. This creates a powerful 'Wow!' spine-tingling ambiance enhanced with the right atmospheric music.

In the center of this environment, is a flying saucer-shaped object on a tall stand about 3 m (10') high. The 'UFO' has led and fiberoptic lighting moving around its edges, with a flying saucer sound effect emanating from it. The stand is made to look like a light beam by surrounding it with backlit layered semi-transparent material. The UFO's housing contains four video projectors pointing at screens in each corner. Its top houses a laser pointer that is pre-programmed to point to objects in the real sky when the roofs are rolled back. The show director or the preprogrammed narrator of the show then explains to the audience what each object is.

Surrounding the UFO is traditional seating with layback backrests for elderly people with back problems or stiffness, so they can recline back easily to look upward when the show takes the audience high into the sky. These seats can be rotated left and right up to 60° to allow patrons sitting in them to see things behind them.

Most agile people will choose to lie on the exciting circular 'flotation' mattress that surrounds the perimeter

A perspective view of a model of a Cosmos Center attraction where the roofs have been removed to show internal details. It incorporates a main entry statement, a Reception Center, a Themed Merchandise Shop, an Administration Center, an Observatory with a large telescope, a Space Theater, a Stargazer's Garden with many moderate-sized telescopes, a large binocular Chair Court for tours of the galaxy, a Mythology Theater, a Space Themed Restaurant, a indoor/outdoor Coffee Lounge, an Amenities Block, an Astronomy Museum, an Interpretative Center, and a Space Train people mover. Model by Dreamtech Designs

of the environment. In a small venue, this would be a series of ordinary mattresses, but for a serious commercial attraction, super-strong waterbeds could be used. They would have a thick foam overlay. This is covered in a tie-down, bright colored commercial, stretchy upholstery cover with fiberoptic lighting underneath it that creates a captivating, mood-setting twinkling starfield appearance. The audience feels like they are floating in the stars. In winter, the waterbeds are heated to give a very pleasing and much-appreciated warmth. Guests are excited by the innovation of lying on a wave surface. For cold locations, guests can choose to wear a poncho specially made of a foil thermal emergency blanket that has the company's logo printed on it. This keeps patrons especially warm when viewing the real night sky while lying down. The ponchos could also be purchased for those to wear when patrons are outside looking through telescopes.

The show commences when everyone is reclining. Very spacey music starts to play as the artificial sky roof parts in the middle and rolls back to reveal the real night sky. This is a very stimulating moment filled with expectation. The show director takes his audience on a mind-expanding 'ride' across our galaxy pointing out constellations, special stars and their colors, the Milky Way and its dust lanes, deep sky objects, the planets, and the Moon and comets when visible. When pointing to each object, images or video of them slowly brightens on the screens in each corner and then they fade away, so as to not diminish the audience's night vision. Appropriate stimulating music should be scored for each object. A computer control system would allow the director to select any object. The computer allows for the time of year and the time of night. If scattered cloud covers a particular object, the show director can select another object that is visible from those on his monitor screen. The Space Lounge Show could last between 12 to 25 minutes.

The venue could operate during the day by showing movies about astronomical subjects and having the director answer questions from the audience about the universe. Many people visiting this venue during the day would be likely to come back for the night show. And those visiting in one season would be enthused to come back in another season to see different objects in other parts of the sky.

The building for this attraction could be small to hold as little as 40 patrons for a small family business operation, or as many as 200 or more for a large commercial company venture. It could hold many more by having a double ring of flotation beds around the perimeter. It is best located at a site that has a reasonably dark sky. The building could be earth integrated by having the exterior walls mounded up with soil over a 1.5 m (5') deep to retain summer warmth in winter, and in summer, the walls will remain cool from the soil storing winter coldness in the soil. This earth could be planted with appealing ground covers that do not grow higher than the top of the walls of the building.

The model of the Space Lounge building below illustrates how the roofs roll away when the night sky viewing shows begins.

The roof is designed as a motorized, dual, roll-away roof, as detailed earlier in this chapter for observatories. In this case, the roof would be arched and lined on the inside. The roller system should be very quiet. The motor should be able to close the roof quickly in the event of an unexpected shower. At times of inclement weather, the director would present a show about some aspect of astronomy with the roof closed.

This concept for a Cosmos Center has special effects projectors lighting the sky for 5 minutes between each changeover of shows. Its buildings feature dome structures with earth-integrated walls for climate control. Concept illustration by Gregg Thompson, Gary Collett

The entry statement has large, internally illuminated models of the planets at the entrance. The Center's signage is lit by fiberoptics.
Concept illustration by Gregg Thompson, Gary Collett

This edu-tainment attraction designed for naked eye observation of the night sky. It is named the 'Space Lounge'. Patrons relax on floatation beds to gaze into the heavens. Their unique experience enlightens them about the wonders of our universe. Concept illustration by Gregg Thompson, Gary Collett

In the Star Garden, visitors observe objects in large telescopes. While waiting their turn, they watch videos that provide details about each object that they are about to observe. Concept illustration by Gregg Thompson, Gary Collett

Left: **A space-themed restaurant with a view across the icescapes of Europa - or the volcanic landscapes of Io – both moons orbiting Jupiter.** Concept illustration by Gregg Thompson, Gary Collett

Right: **The Mythology Theater concept utilizes an artificially created nighttime setting in the round in which actors portray stories about colorful myths from ancient peoples that attempted to explain the workings of the heavens. The audience sits around an artificial but realistic looking fireplace under a realistic-looking, artificial night sky, in which special effects appear to support the storytelling.** Concept illustration by Gregg Thompson, Gary Collett

In summary, dual roof or split shed designs are exceptionally practical as they cater for many needs. They are relatively inexpensive and relatively easy to build.

To increase sales, such a Cosmos Center could also incorporate a space train ride that winds around the entire center, space-themed sweets in mini outdoor shops, space ice-cream stands or trolleys, and sales of themed toys that light up in the night. There could also be a site outside the attraction for campers to pitch their tents and caravans, as well as permanent tents on platforms for rent. People could then stay a night or two to observe the night sky and to rent telescopes.

It could also include computer controlled binocular chairs mentioned in Chapter 8, page 134.

This large concept was developed for a Cosmos Center that would be a substantial, unique, edutainment attraction for a country town on a major highway in a relatively dark location.

I have also developed other concepts for a major attraction themed around astronomy at a dark sky site that requires a much lower budget, but which is nevertheless most impressive due to its uniqueness.

COMMERCIAL ASTROPHOTOGRAPHY OBSERVATORIES

Advanced astrophotographers are renting the use of automated large aperture telescopes at remote dark sky observing sights in the USA, Australia, South Africa, and South America. Amateurs can login to these telescopes and apply the required filters to photograph a list of objects. When completed the data files are accessed via the Internet for them to process their images at home.

The iTelescope network has many automated telescopes housed in a split, roll-off roof observatories located at dark sky sites in California, Australia, New Mexico, and Spain. These telescopes an be hired for use by advanced amateur astrophotographers to take high-quality images under optimal observing conditions at

The remotely controlled telescopes available for hire at the iTelescope observatory at Siding Spring mountain in Australia.

an affordable fee. This saves the cost and time associated with transporting a telescope to a dark site, then setting up and packing up, if one can't afford the high cost of building a private observatory.

These observatories at Fairborn Observatory at Mt Hopkins, have fold back roofs. They house remote-controlled 0.7 m (28") telescopes. Robotic observatories are now appearing in dark sky sites in many parts of the world.

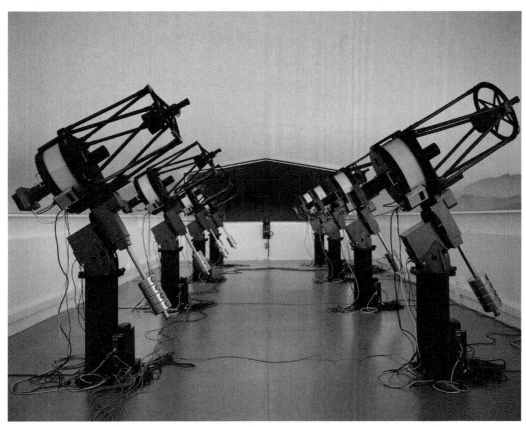

These robotic remote-controlled 400 mm MEarth-South telescopes are located under dark Chilean skies. They search for exo-planets transiting the closest red dwarf stars. They are housed in a large, roll-off roof observatory.

The **Astro Haven observatory** with its fold-back, clamshell roof sections is a relatively new, expensive option for a small observatory. The roof shells are motorized. This product is mostly used for commercial photographic observatories that house remotely controlled, robotic telescopes that require an open sky. For visual observers, they only provide partial protection from wind and stray light when it is coming from the direction directly behind the main body of a shell providing it is left up. Wind coming from the direction of the axis around which the roofs rotate, blows straight through, and stray light will also enter here. One has to get in and out is through a very awkward, small opening under the shells. Only agile people can handle this. It's easier to have steps on the outside near the rotation axis of the shells, and a step on the inside. This design is only really practical as a means of weather protection for an automated photographic observatory where cost is not a problem. I wonder whether sunlight will make the exterior deteriorate over time.

The Astro Haven clamshell observatory

In the next chapter, we look at the largest professional observatories in the world, and the telescopes that they house. Amateur astronomers are becoming increasingly interested in visiting these futuristic facilities when they travel.

Chapter 13
THE WORLD'S LARGEST OBSERVATORIES

The European Southern Observatory (ESO) at Paranal in Chile

Here we see a spectacular panoramic view of one of the world's most advanced telescopes, the Very Large Telescope (the VLT). These telescopes are housed in moveable auxiliary observatories that feed starlight to the main telescope's interferometer to greatly increase the effective aperture of the whole optical system. This results in greater resolution than is possible using one telescope. The telescopes can be moved apart to maximize their combined resolving power. Note that there is an eclipse of the Moon occurring above the VLT observatories.

Once stargazers become familiar with using amateur telescopes, many become excited about visiting some of the world's major observatories when they travel. These observatories house the world's largest telescopes. Most are located in spectacular mountain landscapes, so they are well worth visiting, despite them often being in remote locations. The Internet provides details of how to get to each observatory, what telescopes can be visited, and when tours are available. Check these details as seasons and weather conditions can cause changes to itineraries from time to time.

When visiting these observatories, many amateur astronomers are surprised to learn how much professional astronomical research has developed through utilizing highly innovative technology. This is greatly expanding our knowledge of the universe. When we understand the discoveries that these leading-edge observatories are making, it is easy to appreciate why governments fund these expensive, super-hi-tech telescopes.

THE WORLD'S LARGEST OPTICAL TELESCOPES

The following observatories feature the world's largest telescopes by country location in alphabetical order. The coordinates for longitude and latitude are given as well as the elevation for each observatory. Due to space limitations, only the largest telescopes at each observatory are listed. Most observatories also have many other telescopes.

AUSTRALIA

SIDING SPRING – THE ANGLO-AUSTRALIAN OBSERVATORY (AAO)

It is located near Coonabarabran in New South Wales at 31° 17'S; 149° 04'E. It sits at 1,160 m (3,820') on top of one of a number of impressive remnants of the cores of ancient volcanoes that formed the Warrumbungle Ranges National Park. The observatory has an Interpretive Centre. Along the road up to the AAT there are a number of

Siding Spring Observatory is perched on top of the Warrumbungle Range. Credit: AAT

privately-owned observatories that have nightly stargazing activities for visitors, and there are also observatories with remote controlled telescope that people anywhere in the world can log into to take images of objects and events in the universe. Options for visiting the AAT and the region are available on the Internet. Along the highway to Siding Spring, the local council has created the world's largest scaled down model of the solar system. The AAT dome forms the Sun and the planets are shown on huge 3D billboards along five main roads leading to the observatory.

The largest telescope is the 3.9 M (150") ANGLO-AUSTRALIAN TELESCOPE (AAT).

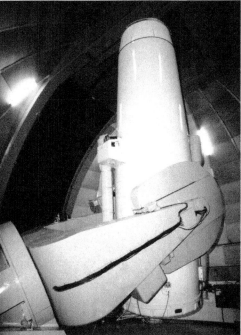

Left: **The AAT with two observers on the balcony on the right for scale. Visitors can look at the 3.9 m telescope from the 4th-floor viewing gallery.**

Right: **The 1.2 M (48") UK SCHMIDT TELESCOPE.**

THE CANARY ISLANDS

THE OBSERVATORIO DEL ROGUE DE LOS MUCHACHOS

This observatory is located on the edge of the volcano near at La Palma Is near Santa Cruz. It is at an altitude of 2,400 m (7,680') at 28° 46'N; 17° 53'W. The combination of the geographical location and climate here causes clouds to form well below the observatories permitting cloud-free skies most of the time. The 50 cm Mons telescope is available for amateurs to use. Bookings must be made beforehand. Visits to the Teide Observatory are suspended between November and March for meteorological reasons.

The Teide Observatory on the Island of Tenerife in the Canary Islands is at an altitude of 2,390m (7,480'). It is on the way to the Teide volcano peak. It houses a number of solar telescopes owned by different countries.

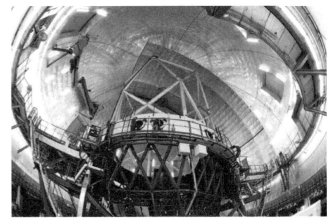

The largest telescope is the 10.4 M (410") GRAN TELESCOPIO CANARIAS, the GTC (The Grand Telescope of the Canary Is). The GTC utilizes a segmented mirror. Credit: Observatorio del Rogue de los Muchachos

The largest nocturnal telescope is the 1500 mm (60") diameter Carlos Sanchez Infrared Telescope. Daytime guided tours are available.

The La Palma Observatories Credit: Observatorio del Rogue de los Muchachos

Here we see a Teide observatory using a laser beam. The el Teide volcano is in the background. Credit: Observatorio del Rogue de los Muchachos

Left: **The 10.2 M (400") WILLIAM HERSCHEL TELESCOPE** Credit: Observatorio del Rogue de los Muchachos

Right: **The 2.54 M(100") ISAAC NEWTON TELESCOPE**

Left: **The Isaac Newton Telescope observatory sits high above the clouds.** Credit: Observatorio del Rogue de los Muchachos

Right: **The SWEDISH 1 M (39") SOLAR TELESCOPE - the world's highest resolution telescope for observing the Sun.**

CHILE

THE EUROPEAN SOUTHERN OBSERVATORY (ESO)

This observatory is described as 'the pre-eminent intergovernmental science and technology organization in astronomy'. ESO plays a leading role in promoting and organizing cooperation in astronomical research. ESO operates three unique world-class observing sites in the Atacama Desert region of Chile. They are at La Silla, Paranal, and Chajnantor.

The La Silla Observatory is at an altitude of 2,400 m (7,870'). It is equipped with several large optical telescopes. Credit: ESO

Left: **largest telescope here is 3.6 m (142") HARPS, (High Accuracy Radial Velocity Planet Searcher),** a spectrograph with unrivaled precision. It is the world's foremost extrasolar planet hunter. Credit: ESO

Right: **The 3.6 m Observatory.** Credit: ESO

Cerro Paranal Observatory with the VLT. Credit: ESO

The Paranal Observatory is 2,600 m (8,500') high and it is situated in one of the driest regions on Earth. It is home to the Very Large Telescope array (VLT) which is the flagship facility of European astronomy. The VLT is an array of four individual telescopes that have main mirrors 25 m (984") in diameter. The VLT also has four additional 1.8 m (70") movable Auxiliary Telescopes. In this interferometric mode, the telescope has a vision as sharp as that of a telescope the size of the separation between the most distant mirrors, which is 200 meters. Guided tours for tourists and the public are offered every Saturday.

The VERY LARGE TELESCOPE, (VLT) is an array of four 8 m (315") telescopes that can work together as one huge telescope with the light gathering capacity of a 16 m (630") telescope. Credit: ESO

The VLTs 1.8 m auxiliary telescopes that act as interferometers. Credit: ESO

Left: One of the four Very Large Telescopes. Credit: ESO

Right: **This is the VISTA (Visible and Infrared Survey Telescope) at Cerro Paranal. It is the world's largest sky mapping survey telescope with a very short focal length mirror 4.1 m (161") in diameter. This telescope's very sensitive camera is revealing a completely new view of the southern sky.** Credit: ESO

The Extremely Large Telescope (ELT) located on top of Cerro Armazones in the Atacama Desert has a 39.3 m (1,547") primary mirror. When completed in 2024, it will be the world's largest optical/infrared telescope. It will produce images 16 times sharper than the HST. Credit: ESO

Left: **THE CERRO PACHON OBSERVATORY** is located at 30° 20S; 70° 59'W is at an altitude of 2,740 m (8,980'). It is home to THE 8.1 M (320") **GEMINI SOUTH TELESCOPE**, which is a twin to the Gemini North telescope on Mauna Kea. It uses a combination of multiple lasers and deformable mirrors to remove atmospheric distortions from its images. This makes the telescope 10 to 20 times more efficient thereby producing razor-sharp images that can detect remarkable detail. Credit: G Snow

Right: The **LAS CAMPANAS OBSERVATORY** situated in the southern Atacama Desert, is located approximately 100 km (62 mi) northeast of the city of La Serena at 29° 00'S; 4° 42'W. It is at an altitude is 2,280 m (7,480'). The **DUAL 6.5M MAGELLAN TELESCOPES** are located here. They are about 60 m (200') apart. One is named after the astronomer Walter Baade, and the other after the philanthropist Landon Clay.

MAUNA KEA

Mauna Kea Observatory is located at 28° 46'N; 17° 53'W on top of an extinct volcano on the Big Island of Hawaii. It is possibly the most impressive location for a major observatory. Its name means 'white mountain' due to it being covered in snow for about half of the year. This is due to it being the highest observatory in the world at an altitude of 4,000 m (14,000'). Measured from its deep ocean base, Mauna Kea is higher than Mt Everest when it is measured from above sea level.

Mauna Kea's summit is above 40% of the Earth's atmosphere, so oxygen levels are considerably reduced. At the 3,000 m (9,000') marker, there is an acclimatization motel for astronomers. They spend a few days to a week there to adjust to the lower levels of oxygen before working on the summit. This gives their body time to make more red blood cells to carry more oxygen. Immediately above this point, vegetation abruptly disappears, and the landscape becomes barren and surreal. You feel like you have been transported to another planet.

Here we see Mauna Kea with it snow-capped top as seen from sea level. Its observatories are just visible protruding from the very top.
Credit: David Yee

Despite the observatory's location being just 20° north of the equator, its high altitude causes it to experience snowstorms in winter. At a latitude of 20°, there is only two seasons, summer and winter. Credit: Robert Richardson

Left: Mauna Kea's summit consists of numerous cinder cones - mini volcanic vents. Their sides have an angle of 45°, being the roll angle of rocks spewed from the vents when they were active. Ash spewed from these vents when the volcano was last active 4,500 years ago. The ground is covered in red, orange, yellow, brown, and black small fractured stones.

Right: The summit of Mauna Kea resembles Mars. It has many volcanic crater dykes. It sits high above cumulous clouds, so much of the time, there is an ocean of cloud lying about a thousand meters (3,000') below the summit. There is no vegetation above 3,000 m (9,000').

Alex Mukensnable was able to capture this amazing image of a partially eclipsed Full Moon rising in the shadow of the tip of Mauna Kea at sunset!

While driving to the top of Mauna Kea, my friends and I were shocked by a very loud explosion that sounded like a gunshot, or possibly the blow out of a tire, but we felt nothing. After stopping and looking for what it was, we found the culprit. In the back of the cruiser, there was a large foil bag of potato flakes that had air pressure suitable for sea level. Under Mauna Kea's much lower air pressure, the air inside the bag expanded so much that it exploded!

Due to the low level of oxygen at this altitude, motor vehicles become sluggish, and their normally invisible exhaust gases become dark, or colored, due to the engine not being able to get enough oxygen to burn all the fuel. At the summit, everything moves slowly as if in slow motion - earthmoving equipment, trucks, cranes, cars, and even people. At that altitude, you need to walk slowly and not exercise too much otherwise, you are likely to experience altitude sickness. This can cause a headache, nausea, a loss of control over one's bowel and bladder, as well as an inability to think clearly.

THE OBSERVATORIES

An aerial view of the summit of Mauna Kea - the highest observatory in the world. Credit: Richard Wainscoat

Mauna Kea's two largest instruments are THE TWIN KECK 10 M (400") TELESCOPES that have segmented mirrors. They can work independently or together as an interferometer to increases the effective size of the telescope's resolving power. When working together they have an effective aperture of 22.8 m (897").

The twin observatories for the Keck telescopes. Credit: Keck Obs

Left: **THE SUBARU OBSERVATORY has an unusual cylindrical 'dome'.**

Right: **THE SUBURA TELESCOPE has an aperture of 8.2 m (323"). It is a single mirror telescope and it is operated by Japan.**

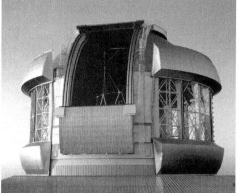

Left: **THE 8.1 M (318") GEMINI NORTH TELESCOPE is a single mirror telescope.** Its twin is located in the southern hemisphere at Cerro Pachon in Chile. The observatory (right) for this telescope has walls that can open up to ventilate the telescope for better seeing.

THE 3.58 M (141") FRENCH-CANADIAN TELESCOPE is a fork mounted prime focus cassegrain.

RUSSIA

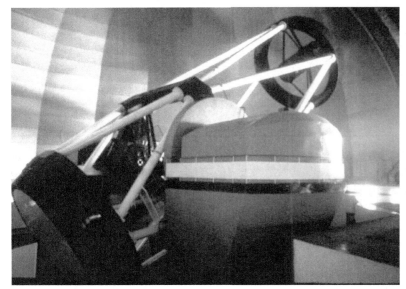

THE 6 M (236") BOLSHOI LARGE AZIMUTH TELESCOPE is at Nizhny, Arkhyz 43° 39'N; 41° 26'E at an altitude of 2,070 m (6,790').

SOUTH AFRICA

THE SOUTH AFRICAN ASTRONOMICAL OBSERVATORY is located at 32° 23S; 20° 49E on an elevated plain 4 hour's drive out of Cape Town. It is at an altitude of 1,760 m (5,770').

Left. **The South African Astronomical Observatory**

Right: **The Observatory of the 9.2 M telescope**

THE 9.2 M (362") SOUTH AFRICAN LARGE TELESCOPE (SALT). There are guided tours from Monday to Saturday. Bookings are essential.

UNITED STATES OF AMERICA

Mt Graham Observatory

The MT GRAHAM OBSERVATORY is at 32° 42'N, 109° 53'W in Arizona. Its altitude is 3,170 m (10,400'). It has the VERITAS TELESCOPES which consists of four 12 m (40") multi-segment mirrors that work together as interferometers. Weather permitting, weekend tours of the Mount Graham Observatory are conducted between mid-May and October.

Mt Hopkins Observatory

MT HOPKINS OBSERVATORY is located in the Santa Rita Mountains, 55 km south of Tucson, Arizona. Its altitude is 2,600m (8,530'). It is home to THE 6.5 M (256") MAGNUM MIRROR TELESCOPE (MMT). The Observatory conducts guided tours for the general public on Mondays, Wednesdays, and Fridays from mid-March through November.

McDonald Observatory

Left: MCDONALD OBSERVATORY is located in the Davis Mountains 725 km (450 m) west of Austin, Texas at 30° 40'N, 104° 00'W. Its altitude is 2,076 m (6,650').

Right: It has the 9.2 M (30') HOBBY-EBERLY TELESCOPE. Its primary mirror consists of 91 hexagonal elements. The Star Party program allows visitors to stargaze through some large telescopes on the mountain.

Kitt Peak Observatory

KITT PEAK NATIONAL OBSERVATORY (KPNO) is 144 km (90 mi) outside Tucson, Arizona is at 31° 17S; 149° 04E. Its altitude is 2,100 m (6,890'). This observatory has a diverse array of large instruments situated on an impressive mountaintop. There are three 1 hour, fee-based guided walking tours per day.

Kitt peak's largest telescope, the 3.8 M (150") MAYALL TELESCOPE. Its Observatory is the one at the rear (right). Credit: Fred Bailey

Left: **Kitt' Peak's WIYN 3.5 m (138") telescope** Right: **and The McMath-Pierce Solar Telescope**

Mt Palomar Observatory

MT PALOMAR in California is at 33° 21'N; 116° 52'W at an altitude of 1,900 m (6,200')

Mt Palomar is home to the historic 5.08 M (200") HALE TELESCOPE was built in 1948. It became the largest telescope in the world until the Russians built an imperfect 6m (236") telescope in 1974. It was not until 1993 that the Twin Keck 400" telescopes were built that it was no longer the best telescope in the world. The observatory is open daily from 9 am to 3 pm, except for Dec 24-25, and during extreme weather conditions.

Mt Wilson Observatory

Mt Wilson Observatory is at 34° 13'N; 118° 04'W at an altitude of 1,740 m (5,700'). It overlooks Los Angeles.

Mt Wilson is home to the famous 0.25 M, 100" HOOKER TELESCOPE, which is historic due to it being used by Edwin Hubble, Milton Humason, and Vesto Slipher to prove that galaxies were island universes like the Milky Way. This telescope was the first to show that the universe was expanding. There are weekend guided tours from April 1 to November 30 from 10 am - 5 pm.

FUTURE GIANT OPTICAL TELESCOPES

There are a number of very large land-based telescopes being built in Chile and on Mauna Kea. There is a laughable trend by some telescope teams to outdo others who are building large telescopes by giving their telescopes sensible names. They give them names such as, the very large telescope, the grand telescope, the extremely large telescope, the overwhelmingly large telescope. Such thoughtless names mean nothing when the next even larger telescope is built. When you see what these telescopes can do, it is hard to believe that the technology is progressing so incredibly fast.

THE LARGE SYNOPTIC SURVEY TELESCOPE (LSST)

The LSST will have a very short focal length primary mirror. It's super wide-angle and super sensitive, digital camera lens will image a huge 7° of the sky in one shot! That's equal to 3,000 images taken with the HST! The LSST's camera lens is 1.6 m (63") across. One image has 3.2 billion pixels, so it would require 1,500 HD TV screens to see a single image in full definition! It will be so fast that it can take a new picture every 20 seconds. At this speed in the clear skies of the Atacama Desert, it will be able to cover the whole sky every 3 nights! This will permit it to see changes across the solar system, our galaxy, and the deep universe as if we are watching a time-lapse movie. It will detect 40 billion stars in our Milky Way, many of which we cannot currently see, and it will also monitor 10 million galaxies. Over a decade, it will discover 1.5 million supernovas, as they explode!

This telescope's software is as important as the telescope's optics and engineering because it can process a massive amount of data immediately! New data processing programs had to be invented to manage this extraordinary amount of big data. The LSST will gather so much data that it will be likely to determine which model of the expanding universe is consistent with observational data.

The LSST will record in real-time the orbits of hundreds of thousands of asteroids and comets moving through the solar system. Because the LSST can record both very faint bodies as well as fast-moving ones, it will not only detect millions of asteroids in the inner solar system, but also those in the Outer Asteroid Belt beyond Neptune, and many out to the inner edge of the Oort Cloud. The LSST will find many dwarf planets, and maybe even other planets in the far reaches of our solar system. It will also detect small asteroids that could collide with Earth in the future.

With the LSST, we will no longer see static images of the sky: we will see the universe changing and evolving before our eyes! The LSST is expected to be operational by 2020. This telescope will be so far beyond any existing telescope, that is hard to believe that something this advanced could exist in our time. It makes us wonder what the next giant leap forward in telescope engineering will be in the decade after that.

Left: **Illustration of the LSST OBSERVATORY**

Right: **Illustration of the 8.4M (27') LSST telescope that will revolutionize astronomy.**

THE GIANT MAGELLAN TELESCOPE (GMT)

The GMT will have 10 times the resolution of the HST. It may even be able to detect chemical signs of life on planets beyond our solar system. It will be able to record exceedingly faint objects, as well as bright ones in great clarity. It will stand 43 m (140') high – comparable to the height of the Statue of Liberty. It is so large that it requires an observatory 22 stories high to house it! The observatory will weigh 2,000 tons and it will rotate with the telescope. The GMT will be located in Chile at the Las Campanas Observatory. It is expected to be fully operational by the mid-2020s.

The Giant Magellan Telescope (GMT) comprises 7 huge mirrors each 8.3 m (327") in diameter. Collectively they will form a mirror 24.3 m (80' or 23,270") across. Note how small the person is compared to the size of the telescope. Credit: Giant Magellan Telescope — GMTO Corporation

THE 30 M EXTREMELY LARGE TELESCOPE

The Thirty Meter Telescope (TMT), also known as the Extremely Large Telescope (ELT). It will be built on Mauna Kea. It will consist of almost 500 segmented mirrors with 9 times the resolving power of the twin Keck telescopes. It is designed to cover the spectrum from near ultraviolet, through visible light, to the mid-infrared. It is expected to be operational by the mid-2020s.

COMPARISON OF PRIMARY MIRROR DIAMETERS ON LARGE TELESCOPES

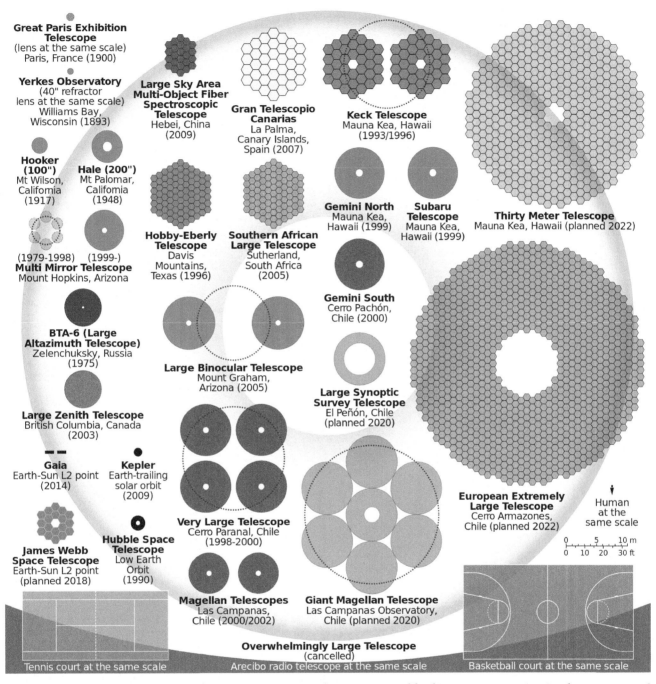

When you see the relative sizes of large telescope primary mirror diameters set out like this, you can appreciate just how enormous the latest generation of telescopes is. When you think about the sizes of a tennis court or a basketball court, this helps you relate to the sizes of the largest telescopes. It won't be long before optical telescopes the size of the proposed Overwhelmingly Large Telescope are built. Despite how small the HST mirror is in comparison to the large ground-based telescopes, it can outperform all but the very largest telescopes now being built, so we need to have telescopes larger than the HST in space, or on the Moon. Space telescopes that do not have to cope with the blurring effect of our atmosphere, are far more effective than ground-based telescopes, however, the cost of getting them into orbit is very expensive.

SIZE COMPARISON OF THE WORLD'S LARGEST TELESCOPES

Very Large Telescope

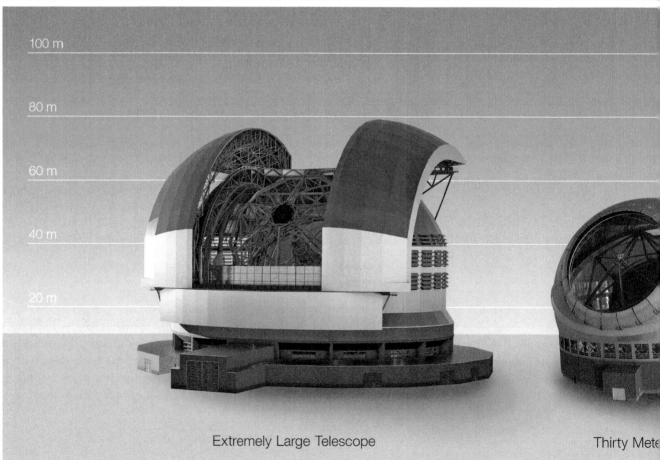

Extremely Large Telescope

Thirty Mete[r]

Illustrations: **ESO**

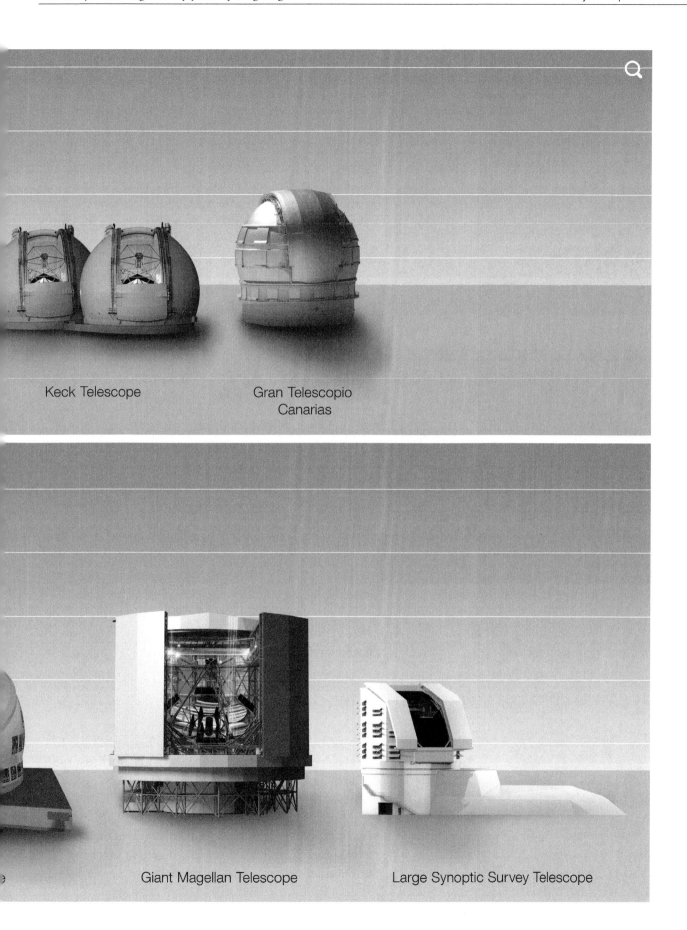

THE LARGEST OPTICAL TELESCOPES IN SPACE

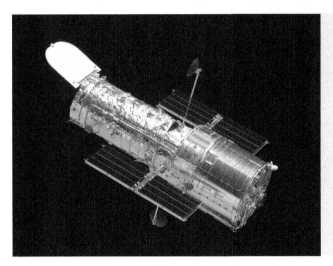

Hubble Space Telescope (HST)

THE HUBBLE SPACE TELESCOPE (HST) has a 2.4 m (94") main mirror. It is located at an altitude of 570 km (350 m). The HST orbits Earth every 97 minutes at 28,000 kph (17,500 mph). The HST operates on solar power. It is the most outstanding, visible light, optical telescope so far built. It cost US$1.5B. When it passes over your location, it is visible as a bright, moving, star-like satellite. Very disappointingly, when it has a major fault, it will not be possible to repair it now that NASA no longer has any space shuttles.

Spitzer Space Telescope

THE SPITZER SPACE TELESCOPE was launched in 2003. It is an infrared telescope that measures the heat given off by objects throughout the universe. It can look inside interstellar dust clouds to see solar systems evolving, and it can see the heat from dark objects, as well as matter being ejected from objects such as supermassive black holes in the centers of active galaxies, known as quasars.

Herschel Space Telescope

THE HERSCHEL SPACE TELESCOPE will observe the universe in the far infrared and sub-millimeter wavelengths from 55 to 672 µm. This is the part of the spectrum in which much of the universe radiates energy. It will be able to observe objects that other telescopes cannot see such as those obscured by dust and objects that are very cold. It is expected to operate for a few years. It will orbit the Sun 1.5 million km from Earth at one of its La Grangian points (These are 5 locations in an orbit around the Earth where the orbital motion of a spacecraft interacts with the gravity of the Earth, the Moon and the Sun in a way that makes this orbit stable.)

Kepler Space Telescope

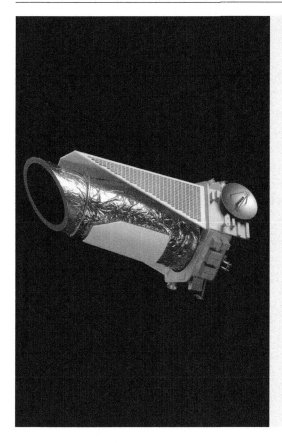

THE KEPLER SPACE TELESCOPE studies stars in a limited region of the sky 100 square degrees in diameter in the constellation Cygnus near the star Vega. It detects solar systems where the orbits of their planets are aligned with our line of sight so that their planets pass in front of their star. Less than 1 in 100 stars have planets aligned in this way. Kepler will observe hundreds of thousands of stars to have a statistical chance of discovering many planets using this method. When a planet is found to be eclipsing its star, it dims the star ever so slightly on each orbit. Kepler needs to see multiple transits of each planet across its star to ensure that the dimming is regular, and of the same magnitude each time in order to ensure it is a planet. It can only detect planets that orbit their star quickly because they are close to it. Kepler can detect the approximate size of a planet and how far away it is from its star. Most planets that Kepler detects are larger than Earth. Many are larger than Jupiter, and some of these orbit very close to their star. In 2015, Kepler found its first Earth-like planet, which is about 60% larger than the Earth. It is now finding many more orbiting their star in the habitable zone. Most stars that Kepler observes are 500 to 2,000 light years away. It has already discovered thousands of exo-planets.

James Webb Space Telescope (JWST)

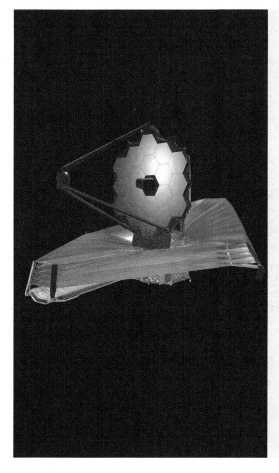

THE JAMES WEBB SPACE TELESCOPE (JWST) is named after the Director of NASA between 1961 to 1968. He envisaged a space telescope with great resolving power. The JWST will photograph the universe in visible light and into the far infrared. Due to the high recessional velocities of distant galaxies, their wavelengths are highly red-shifted, thereby pushing their light from the high energy UV and the optical parts of the spectrum, into the near infrared. The JWST will observe the universe in wavelengths from 0.6 to 28.5 microns in high resolution using its large 6.5 m (256") primary mirror, which is far larger than the HST's 2.4 m mirror. The JWST will be able to observe from 0.6 to 28 microns. It will be able to observe the first galaxies to form in the universe. With its wide field of view, it will be able to cover more than 15 times the area of sky that the HST can. And it will deliver better spatial resolution than the Spitzer Space Telescope. The JWST's heat shield will be the size of a tennis court, so it will be a very bright satellite. It will be used to observe the formation of stars and the first galaxies, so we can see how they formed. It will tell us the physical and chemical properties of other solar systems and our own, and it will help us estimate the potential for life elsewhere in our solar system and galaxy. The JWST can observe 100 objects simultaneously!

WFIRST (Wide Field Infrared Survey Telescope)

WFIRST (Wide Field Infrared Survey Telescope) will be launched in 2020. It has a 2.4 m (95") main mirror. It will have the same resolution as the HST, yet it will be able to cover 100 times as much sky in each exposure! It will capture a million galaxies in just one image. WFIRST will study dark energy, exo-planets, and the acceleration of the expansion of the universe. It has a coronagraph, which will dim the light of a star to make its planets up to a billion times brighter!

StarShade Planet-finding Telescope

The proposed STARSHADE PLANET-FINDING TELESCOPE will be able to directly image planets orbiting other stars. The telescopes designers expect that all stars will have planets and that one in five will have Earth-like planets that may support life. At present, astronomers find planets by their gravitational tug on their star, or by the dimming of the light of a star when a planet passes in front of it. At present, we cannot see the planets directly due to their extreme faintness against the glare of the star. However, the StarShade telescope will overcome this limitation by using a flower-like star shade that has ingenious, critically shaped petals that can eliminate most of a star's glare, and it will greatly reduce the diffraction of light. This will make the star 10 billion times dimmer, allowing planets to be seen directly! The star shade has to be positioned 50,000 km away from its space telescope. The extraordinary technology that is being developed now will allow this telescope to achieve its ambitious goals. (See the TED talk on YouTube to see how the technology will work.)

THE WORLD'S LARGEST RADIO TELESCOPES

ARECIBO RADIO TELESCOPE

THE ARECIBO RADIO TELESCOPE in Puerto Rico is 305 m (1,000') in diameter. It is suspended in a hollow between the surrounding hills. The facility covers 25 acres. The dish shape is achieved by suspending cables from its rim. The receiver is suspended from three towers placed on the hill tops surrounding it. Because the dish is static, the receiver has to be moved so that the telescope can observe different parts of the sky. It is steered by the suspension cables pulling it one way or another. Because of this, it has limited coverage of the sky.

PARKES RADIO TELESCOPE

The 64 m (210') dish of THE PARKES RADIO TELESCOPE is located near the small town of Parkes in New South Wales, Australia. It is a fully steerable dish that is very well engineered for stability and pointing accuracy, so it has high resolution. A 3D theatre on site shows a variety of 30 minute movies about astronomy. Credit: S Amy CSIRO

AUSTRALIA TELESCOPE COMPACT ARRAY

THE AUSTRALIA TELESCOPE COMPACT ARRAY at Narrabri, NSW in Australia is a series of six 22 m (70') dishes that move along tracks to increase their separation to improve resolution. When combined, they have the resolving power of a telescope equal to the distance of their widest separation. The facility has a visitor's center.

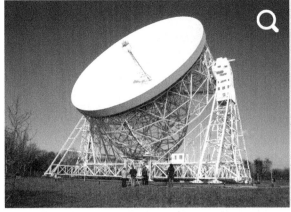

Lovell Radio Telescope

The 76 m (250') LOVELL RADIO TELESCOPE at the JODRELL BANK OBSERVATORY in England is an historic, fully steerable dish. It is famous for picking up the first satellite Sputnik, that orbited the Earth, and for the discovery of pulsars (neutron stars). It has an exhibition center for visitors.

Very Large Array (VLA)

THE VERY LARGE ARRAY (VLA) in New Mexico, USA is the world's premier radio telescope consisting of 27 antennas each 25 m (82') in diameter. They are arranged in a Y shaped layout. The visitor center offers free guided tours from 8.30 am until sunset. The VLA and the Arecibo telescope were featured in the cleverly conceived movie and book titled Contact, written by Carl Sagan.

APEX Millimeter and Submillimeter Telescope

The ESO's Chajnantor site is home to the 12 m APEX millimeter and submillimeter telescope.

ESO's ALMA (Atacama Large Millimeter-submillimeter Array)

The ESO's ALMA (Atacama Large Millimeter-submillimeter Array) combines 66 12m radio telescopes which gives it an effective diameter of 14,000 m (46,000'). It could detect a golf ball 15 km (9 mi) away. It is at an altitude of 5,000 m (16,500'.) The dishes are perfect to 20 millionths of a meter. Each dish weighs 100 tons, despite being made of carbon fiber reinforced plastic. Its detectors are chilled to 2°. It cost $1.3 billion to build. It was funded by North America, Europe, and East Asian countries. Credit: Tafreshi

SKA (Square Kilometre Array)

SKA (THE SQUARE KILOMETRE ARRAY) is being constructed in Western Australia. It will be the world's largest radio telescope. The dishes in its concentrated central core connect to other radio telescopes across the continent to give it an effective diameter of 3,000 km (1,900 mi). Completion is scheduled for 2022.

Aperture Spherical Radio Telescope (FAST)

FAST (The Five hundred meter Aperture Spherical Radio Telescope) (1,600') is located in the Guizhou province in China at 25.6525° 106.8567°E. Like the 300 m Arecibo radio telescope, it is a suspended dish that hangs from six tall pylons surrounding a natural depression. Its spherical surface causes spherical aberration. To compensate for this, the surface plates continually adjust to simulate a parabola to achieve high resolution images. It will be 3 times more sensitive than the Arecibo telescope. Its feed antenna is suspended 140 m above the reflector. It is driven by cables and servomechanisms to allow it to cover 40° of the sky from directly overhead.

Giant Metrewave Radio Telescope

GMRT (THE GIANT METREWAVE RADIO TELESCOPE) in Pune in India consists of 30 fully steerable 45 m parabolic dishes that observe at a 1 meter wavelength.

Chapter 14
PRESERVING THE NIGHT'S NATURAL BEAUTY

Due to high levels of both air and light pollution, no stars can be seen in the night sky above Shanghai.

If there was no air pollution, under a naturally dark night sky as might occur during a blackout, this is what Shanghai would look like.
Credit: Thierry Cohen

In this chapter, we will learn why most people no longer see numerous faint stars, shooting stars, great comets, and the Milky Way – all of which were clearly visible in city suburban areas only 50-60 years ago. We'll investigate what can be done to greatly reduce light pollution that has caused us to no longer see a natural dark sky unless we travel well away from cities to undeveloped country areas.

In the 1960s, most suburban street lighting consisted of no more than a single 150 watt incandescent light bulb suspended over crossroads. On main roads, there were well-spaced incandescent lights on telegraph poles and only near intersections or around tight corners. In most places, the very limited amount of street lighting went out at 11 pm, so after this time, the Milky Way was visible even in inner city suburbs! There was almost no outdoor floodlighting of buildings, car sale yards, outdoor car parks, or sports fields. Homes did not have spotlights, and there was no need for security lighting, or brilliant advertising signage. So, the term 'light pollution' did not exist before the 1970s. There were fewer cars on the roads then, and few people went out on week nights. In those days, cars had headlights, so the huge cost of lighting highways along roads was not deemed necessary.

Once government-owned electricity companies started to be sold off to private enterprise, these utilities then pushed hard to sell as much electricity as possible to the government, business, and residents in order to increase their profits. The manufacturers of outdoor light fittings and those producing light bulbs also got behind the move to sell more electricity to sell more of their product as well to increase their profits also. It soon became common for most forms of outdoor security and advertising lighting to be left on all night, even though this was of almost no value and waste of costly energy.

Utility companies sold governments the concept of installing expensive continuous lighting systems along main roads on the basis that it improved motorist's safety. Instead of it being turned out at 11 pm when there

Once we could see from close at home, sights like this taken in Colorado, but nowadays, to see a naturally dark sky, we are forced to travel well away from developed areas. And even then, there are often light domes around the horizon from distant cities and towns, so there are few places left in developed countries to see a naturally dark sky. This is especially the case in Europe, the USA, and Japan. Credit: Dan Driscose

was little traffic, as had been the case, they convinced governments and city councils to leave them on all night. Lighting along busy, high-speed motorways where there were intersections could be justified, but continuous street lighting along relatively straight main roads was not required and nor was it needed in low speed, low traffic suburban streets. However, governments were bribed to light these roads, and then to leave them on from dusk until dawn! To help push this, the utilities promoted fear of the dark, and the threat of violent attacks even though there were no statistics to support that.

Waste light from cities that is directed skyward reflects off air and moisture molecules thereby lighting up the night sky. However, this light pollution can be greatly reduced through the use of good light fittings that limit the amount of waste light that goes beyond the area of intended use. It is crazy for us to pay a high cost for electricity to illuminate outer space, while at the same time, we lose seeing the beauty of our galaxy! Most city dwellers today have never seen the Milky Way due to light pollution. When there is a blackout and they get to see it, they phone 911 to report smoke across the sky!

Urban sky glow hides our view of the stars.

THE OVERSELLING OF OUTDOOR LIGHTING

By the 1980s, lighting manufacturers and electricity supply companies saw the opportunity to hugely increase their sales and profits by promoting reasons use more outdoor lighting. They encouraged the lighting of billboards and general advertising signage, as well as the illumination of architecture, homes and gardens, parklands, and sports grounds so they could be used at night. And they persuaded governments to massively increase their expenditure on street and freeway lighting. All these things became major sources of light pollution, not through their use, but due to poorly designed light fittings that allowed much of their light to go straight into the sky and surrounding homes.

WASTED LIGHT AND ENERGY

Approximately 40% of all outdoor lighting is waste light. The cost of this waste energy globally is estimated to be around a trillion dollars per year! Most politicians and taxpayers have no idea how costly waste light is. Rapidly rising power costs are increasing the need for an audit of the cost of power for public lighting and how it can be reduced.

Just as the Californian government made car manufacturers fit catalytic converters to car exhausts to stop air pollution, so should governments introduce laws to make lighting manufacturers design light-fitting that do not have waste light. **When old street light fittings are replaced, they must be replaced with environment-friendly fittings that only allow the** light to go only where it is needed. Such legislation would be a simple solution to a large percentage of light pollution. It would be inexpensive and practical. It would also reduce the cost of energy for everyone. No one would be disadvantaged by this.

BILLBOARD AND SIGNAGE LIGHTING

The bulk of lighting for architecture, billboards, and signage is typically lit from below causing much waste light to radiate directly into the night sky to no one's advantage. When lit from below, even light hitting the sign reflects into the sky. Most of this type of lighting could be lit from above to much better effect and it would greatly reduce waste light entering the atmosphere. **Advertising lighting should have shields to restrict the light to only the area of the sign.** There needs to be ordinances to make all outdoor lighting adequately shielded.

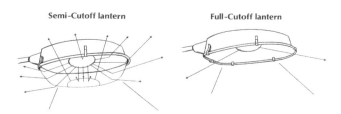

Left: **Semi-cut-off street lights** waste a large percentage of their light by having it radiate far beyond the street into neighboring homes where it contributes to sleeping disorders. As well, around 25% of the waste light from these fittings enters the night sky. They also cause glare for motorists. Because there are so many of these, they are the worst culprit for creating light pollution.

Right: **Full-cut-off street lights** keep most of the light on the road where it is needed, and they greatly reduce glare in motorist's eyes.

Unshielded roadside billboard lighting like this can lose up to 70% of the light beyond the signage area thereby causing significant light pollution of the night sky. Credit: Gregg Thompson

An overabundance of spill light from old styles of unshielded street lighting has cost us our beautiful view of the cosmos, and it wastes billions of dollars globally every year in waste energy. This money could be well spent on replacing these old fittings with ones that limit light pollution.

STREET LIGHTING

For decades, there have been two main types of fixtures for highway lighting:

1. semi-cut-off fittings which have a high percentage of waste light through trying to spread their light too far, and,
2. full-cut-off fittings which produce little waste light.

CUT-OFF STREET LIGHTING PROTECTS THE NIGHT SKY

Cut-off lighting makes pedestrians, animals, or obstacles on the road more visible. This is due to variations in brightness along the road that make both dark and light objects more easily seen.

New shielded LED fixtures waste less light than old bulb fittings, and they use much less electricity. They also require less maintenance. They produce a bright white light that motorists and pedestrians prefer to the cold, bluish Mercury Vapor lights, or the pinkish high-pressure Sodium Vapor ones, and the very unnatural amber low-pressure Sodium Vapor lights. In the city of Davis in California, residents demanded a complete replacement of the new led lights because they were too bright. It is easy to make them less bright. The American Medical Association has warned that if led lighting is too bright and too blue, it can cause harmful effects to human health.

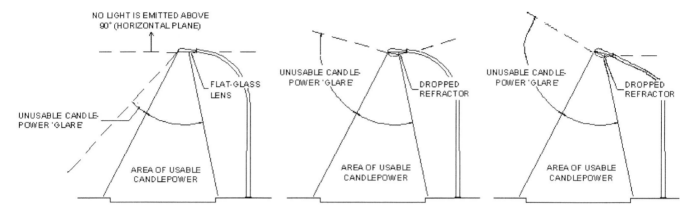

Fig 1. Cut-off light **Fig 2. Semi-cut-off flat** **Fig 3. Semi-cut-off angled up**

The full-cut-off street light fixture in Fig 1 does not allow direct light into the sky, whereas the semi-cut-off fixtures in Figs 2 and 3 do.

Left: **Old yellowish sodium vapor street lighting**

Right: **New blue-white LED lighting**

Two examples of modern LED street lighting that directs all the light downward and only where it is required. It greatly reduces waste light and therefore light pollution, and it also eliminates glare in motorists' eyes.

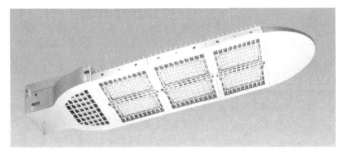 

LED lighting fittings are now replacing the old, poor-quality fittings. They reduce glare and energy costs, and they greatly reduce light pollution of the night sky as well as light trespass into neighboring properties.

From a slightly higher viewing angle than the previous images, and with poles that are not as high, these street lights are barely visible, yet the street is well lit. Note that there is no spill light beyond the footpath. The only light entering the night sky is reflected light from the dark bitumen. Trees block a fair percentage of the light reflected off the road. Credit: General Electric

The LED fittings here produces no waste light beyond their intended area of use i.e. the road. And they create some variation in light to make both dark and light objects on the road easily visible Credit: General Electric

This large area of two levels of roadway is well lit without any direct light from the street lights radiating into the sky. When all old lighting fittings are replaced with fittings like this in the future, light pollution from street lighting will be considerably reduced. Credit: Phillips Lighting

GARDEN AND PARKLAND LIGHTING

In decades gone by, lighting in parks was typically a street light fitting over a path. Its glare made most of the unlit garden difficult to see, and somewhat threatening. The lights spill light in every direction including skyward. There was no attempt at creating pleasant atmospheric lighting because most parks were closed at night unless they were used as a thoroughfare, so no one cared how bad the lighting was. People were not aware of the concept of aesthetic landscape lighting.

Good landscape lighting design does not allow light from a luminaire to go directly into people's eyes and cause glare. Low set, shielded lights that project light downward onto paths and not into pedestrian's eyes, are what is required. Path light sources should emit light from below the eye level of an average height adult.

Light sources should be hidden in many cases so that people are drawn to the illuminated subject, and not the light source. Up-lighting should only be used to light the underside of a dense tree canopy. This way, light does not spill into the night sky. Spot-lighting should light features, such as sculptures or fountains and waterfalls so they contrast against subtle lighting effects throughout the rest of the garden on such things as - tree branches, flowering shrubs, striking plants like giant cacti, ground covers, rock

Overhead down-lighting can be a sculpturistic feature. It can replace traditional path lighting. Spots of subtle light as opposed to one bright area, is far more atmospheric and practical. Up-lighting of the dense foliage of non-deciduous trees is also a good means of indirect lighting.

outcrops, cliffs, sun canopies, garden furniture, textured feature walls, and waterways etc. Colored lighting is a good way of reducing light pollution, as it only emits one color. Fiber optics and led lighting can be used to great effect, and they cause almost no light pollution problems.

Gardens can be very well lit without having to spill light into the atmosphere, or into the living quarters of nearby units.

Left: This is a good example of low-set path lighting that provides plenty of light onto the path and steps without putting glare into walker's eyes.

Right: Up-lighting into a dense tree canopy creates a dramatic effect while allowing its reflected light to softly illuminate a pathway.

Left: Louvered fittings for walkways eliminate glare in pedestrian's eyes by directing all their light downward where it is needed.

Right: Light from a led at the center top of the fitting reflects off a pyramid shape and back into a reflector at the top, which then directs the light onto the pathway and garden.

Left: **Street lights with reflective shades** give a good distribution of light on the ground and they do not allow light to radiate into the sky.

Center and Right: **Shielded spotlights** send the light where it is needed without spill light occurring.

SHIELDED LIGHT FITTINGS FOR BUSINESSES, COMMUNITY CENTERS, AND HOMES

The reason outdoor lighting causes light pollution is because much spill light is directed straight into the night sky or neighboring properties. This occurs due to the light fittings not being shielded so that the light is restricted to where it is needed. **Light shields** are inexpensive, and by concentrating the light where it is needed, a lower wattage light bulb can be used to reduce electricity costs.

Many lights used for illuminating architecture and signage could be turned off after 10 pm or 11 pm when there is little need for them rather than have them wasting electricity all night long.

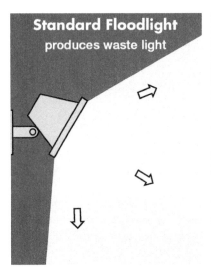
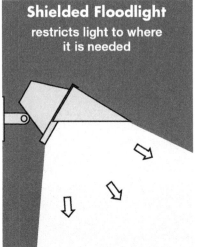

Left: **Unshielded lights permit much waste light to enter the sky and neighboring properties.**

Right: **Shielded lighting keeps the light where it's needed, thereby making it more effective and reducing light trespass that causes friction with neighbors.**

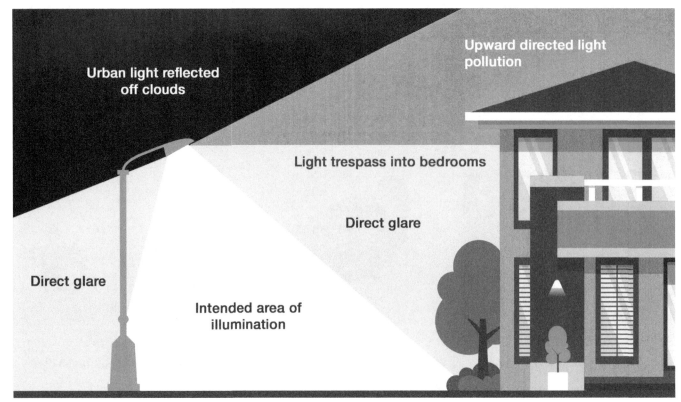

Left: This type of fixture has adjustable shields or baffles that limit the light going beyond where it's needed.

Center and Right: These lights are classic examples of sensor-activated, unshielded outdoor 'security' lights that cause glare and light trespass into neighboring properties. Because they have no shielding, they permit light to travel into the night sky.

Good	Poor	Very poor
High utilance Minimal waste light Minimal sky-glow	Low utilance Considerable waste light Considerable sky-glow	Very low utilance Majority of light wasted Very bad sky-glow
Cutoff	Semi-Cutoff	Conical
Highmast Downlight	Cylindrical	Double Fluoro
Louvred Flood	Unshielded Flood	Spherical
Boxlight	Paraflood	Wall Security

Left column: **Street lights with reflective shades give a good distribution of light on the ground and they do not allow light to radiate into the sky.**

Center and Right columns: **Shielded spotlights send the light where it is needed without spill light occurring.**

SO-CALLED 'SECURITY' LIGHTING

The concept of 'security lighting' was an unknown term until the late 1970s. At this time, the electricity supply companies and the lighting manufacturers realized that if they traded on many people's genetic fear of the dark, they could sell more lighting fixtures and electricity. A fear of the dark is imprinted in many people's memories from very primitive times when humans risked being preyed upon by wild animals in the night. Obviously, this is no longer the case, but in many people, our genes still drive this irrational fear. By using this fear, they could dramatically increase their sales for floodlighting to give people the false impression that the light made them more secure.

They promoted a completely false concern that businesses which had outdoor storage yards could have thieves steal their goods if they did not install 'security' lighting that stayed on all night. 'Security' lighting actually helps thieves because it provides them with a safe and easy environment in which they can steal stock more quickly than if they had to use torchlight and risk being seen as burglars. When thieves wear clothing like a company's work uniform or a security company's uniform, no one pays any attention to them, as people seeing them assume they are the company's staff working back late to get an order out – yes, for themselves!

Security lighting is much more effective if it only comes on briefly when it is activated by a motion sensor. Having a light turn on and off quickly startles an intruder and it attracts attention from passers-by, security guards, and the police so thieves are likely to flee. Plunging intruders into darkness, again and again, can cause them to use flash lights and this makes them more obvious.

During my years as chairman for an association formed to educate government and the public against installing obtrusive outdoor lighting, I learned of a number of cases where security guards were deceived by thieves dressed as staff. This happened to one of our committee members! When security staff checked on the thieves, they told them that they were doing repairs as if they were staff, or getting an urgent shipment ready, so the security guards left them alone. In fact, they were loading up hundreds of thousands of dollars' worth of the company's stock under 'security' lighting! There were companies that were not taken in by the sales pitch for security lighting because they said they had not been broken into in all the years they had been in business, so why would they need it now. Besides, a real security dog, or the sound of one set off by a motion sensor, is the cheapest and most effective deterrent.

When schools installed security lighting, break-ins went up significantly, and so did the number of school buildings that were burnt down. A lit environment made it easy for vandals to steal items and to start fires without having to use torch lights which are more easily noticed. This type of vandalism costs governments, businesses, and homeowners large sums of money. **Contrary to popular belief, statistics shows that 'security' lighting makes crime more likely.**

The electricity industry salesmen targeted shopping centers, and government transportation centers to try to convince them that people would not feel safe, if they too did not increase their lighting. They then encouraged schools and universities to install security lighting to reduce the perceived chance of muggings, theft, and rape. However, studies done in the US found that university students were far more susceptible to physical and sexual attacks when they stood under a street light. By doing so, this made a criminal's target easier to see their build, looks, age, gender, and dress. In low light, it is hard to see anyone at a distance, let alone what they look like. And it is especially difficult to see them if they are not wearing light colored clothes.

If there is a prowler around your home, your best strategy is to turn on outdoor lighting and turn off all indoor lights, so you cannot be seen. Interior lights will make you a more vulnerable target, but darkness will give you the advantage of knowing your territory much better than a potential intruder. In the darkness, you can plant obstacles that will cause an intruder harm. In darkness, you can move around your home much more easily than an intruder can, especially as they do not know your layout. And you can hide easily without an intruder being aware of your location. At your mains box, turn the switches to your interior lights off so an intruder cannot turn lights on to find you, or to rob you.

Lighting gives many people a false feeling of security when, in fact, darkness is one's real security.

ANTI-POLLUTION LAWS

Fortunately, to stop cities being chocked with air pollution, many western states have introduced clean air laws that require:

1. the installation of catalytic converters on motor vehicles to eliminate sooty exhaust fumes,
2. factories and power stations to install filters on their chimneys to eliminate smoke particles and,
3. homeowners to no longer have burn-offs in their backyards, and to no longer burn timber in their fireplaces. These practices produced a lot of smoke particles that light would reflect off.

These laws have greatly reduced the number of pollution particles in the air, so this has helped to minimize light pollution.

Anti-pollution laws were also introduced for water in order to stop polluted run-offs from farms, mines, and rubbish dumps. As well, factories could no longer dump poisonous chemicals and other waste products into rivers and sewerage systems. These Clean Air and Water Acts have hugely reduced both air and water pollution. Today, most major cities in developed Western countries have clean air and water.

Contact:
International Dark-Sky Association
3223 North First Ave, Tucson, AZ 85719
Phone: (520) 293-3198

LIGHT POLLUTION ORDINANCES

The Dark Sky Association in the US educates governments and the public on good outdoor lighting design. Their recommended ordinances for cities and towns greatly reduce the cost of waste energy. These ordinances limit light trespass into neighboring properties, and they greatly reduce unnecessary light pollution of the night sky.

The city of Flagstaff, Arizona was first to introduce an outdoor lighting ordinance over 50 years ago. This was to protect the night sky at the state's major astronomical observatories surrounding Flagstaff. The ordinance restricted searchlight advertising, together with overly bright signage, and unshielded security lighting. It also promoted cut-off street lighting for motorways. Since this was introduced, many other city councils have invoked similar ordinances. Such ordinances should be uniform across the globe.

OBTRUSIVE LIGHT INCREASES HEALTH RISKS

Medical research into sleep deprivation has shown that light trespass into bedrooms can cause sleep loss resulting in depression, a significant drop in one's immunity, memory loss, and an over-production of cortisol, which can lead to cardiovascular problems.

There has also been research conducted that has highlighted other dangerous consequences caused by unshielded outdoor commercial lighting. Under poor weather conditions such as rain, snow, and fog, motorists can become confused by unshielded outdoor lighting. This has caused very serious accidents, a number of which have been fatal.

In poor weather, in the past, airline pilots confused obtrusive lighting around airports for airstrip landing lights. When visibility is low in bad weather conditions, ship's pilots have confused background commercial lights around harbors for channel markers. And similarly, in wet weather, train drivers have had similar trouble confusing unshielded commercial lighting with signals. These cases have led to calamities that caused the loss of life.

ECOLOGICAL DAMAGE FROM WASTE LIGHT

Ecologists report increasingly large numbers of nocturnal animals being adversely affected by increasing light pollution. Newly hatched turtles crawl toward beach resort lights instead of towards the Moon rising over the ocean. This has caused the hatchlings to cross a road where they are killed in large numbers.

Night flying birds that navigate by the stars become disorientated by brilliant lights causing them to die in their thousands. Fireflies and glow-worms cannot

A naturally dark sky is still visible outside Flagstaff due to ordinances that restrict waste light. Credit: Dan & Cindy Duriscoe USNO.

This is a composite image of the Earth at night taken by satellites when the sky was clear over each region. Note the large areas of light pollution. Tens of billions of dollars are wasted each year by artificial light that directly illuminates outer space, and which serves no practical purpose in doing so. If ordinances were introduced to stop this waste light, this would greatly reduce light pollution. We would then not have to travel far from towns and cities to see the beauty of our cosmos. Credit: NASA.

compete with manmade lights to attract mates and prey, so they are becoming extinct. Numerous species of insects are drawn to artificial lights and die in large numbers.

There are also many species of flora that cannot survive having their day/night cycle upset by lights that are on throughout the night. Many life forms have become extinct due to outdoor lighting.

A WORLD AWASH WITH WASTE LIGHT

A powerful example of how waste light affects the globe is illustrated by the following NASA satellite image of the world at night. Numerous cloud-free images were composited to make this whole-world view. It plainly shows that light pollution is extreme across much of the US, particularly in the eastern half. Most of Europe is swamped in light pollution. So is Japan, Eastern China, and SE Asia. Parts of the Middle East have high levels of light pollution around high population areas. It also affects the large population regions in South America, Russia, Eastern Australia, and South Africa. If there was no direct waste light illuminating the night sky, the following image from space would show much less light pollution across the globe.

Let's hope that government lighting ordinances and advances in lighting technology will allow us to be able to see the stars from our suburban homes again in the not too distant future.

In the next chapter, we will look at some of Earth's most interesting astronomical phenomena.

The Milky Way from Monoceros to Perseus. Orion is rising with Sirius above the bush hut. Capella is at the top with the Hyades and Pleiades to the right of upper center.

Hopefully, light pollution may reduce in the future and we will see the stars brightly without having to go far from urban development.

This extraordinary captures the Haleakla Observatory on the rim of the Haleakla volcano on Maui with light pollution emanating from towns along the coast and the Milky Way rising into the sky to the left. The air is so transparent stars go almost the horizon. Credit: Ben Cooper / Launchphotography.com

INDEX

SYMBOLS

3C 273 Quasar— See *Stars*
55 Cancri, Super-Earth Planet— See *Exoplanets*

A

Airglow **90, 102, 115**
Albedo **85**
Aphelion **83**— See *Moon*
Apparent Diameter **72**
Apparent Magnitude **71**
Astro-camp **10**
Astronomical Unit (AU) **73**
Astronomy Clubs **9**
Astronomy Themed Edu-tainment Attractions **219**
Astrophotography **179**
    Auroras **185**
    Comets **185**
    Deep Sky **193**
    Galaxies **198**
    Lunar Detail **189**
    Meteor Showers **186**
    Milky Way **183**
    Moon in Color **188**
    Nightscape **197**
    Planets **192**
    Star Trails **181**
    Timelapse Video **198**
Atmosphere of Earth **93, 94**
    Cloud Types **95**
    Structure of the Atmosphere **93**
        Exosphere **94**
        Ionosphere **93**
        Mesosphere **93**
        Stratosphere **93**
        Thermosphere **93**
        Troposphere **93**
Auroras **42**

B

Binoculars **123**
    3D Binocular Vision **128**
    Binocular Mounts **132**
    Focusing Binoculars **128**
    Rotating Binocular Chair **134**
    Testing Binoculars **136**

C

Cardinal Points **89**
Cataracts **121**
Celestial Poles **75**
Comets
    Comet Tempel-Tuttle **27**
    Halley's Comet **24**
Conjunctions **82**
Constellations **67, 68, 69, 73, 75**
Culmination **81**

D

Dark Adaption **113, 114**— See *Observing*
Declination **77**
Deep Sky Objects **83**
Diurnal Rotation **77**
Drawing Astronomical Objects **156**
    Deep Sky Objects **165**
    Drawing in Color **172**
    Moon **160**
    Night Sky Paintings **175**
    Pencil Drawings **159**
    Planets **161**
    Recording Observations Scientifically **176**
    Sunspots **160**
Dwarf Planets **85**

E

Earth Lights **41**
Ecliptic **81**
Electromagnetic Spectrum **73**
Elongation **82**
Equinoxes **81**

G

Galaxies **16**
    Galaxy Groups and Clusters **16**
    Galaxy Superclusters **17**
Galileo **2**
Gravity **17**

I

Integrated Magnitude **71**
Intrinsic Brightness **72**

L

Life
    Bioluminescence **34**
Lightning Displays **42**
Light Pollution **260**
    Anti-Pollution Laws **273**
    Ecological Damage **273**
    Light Pollution Ordinances **273**
    Outdoor Lighting **268**
    Security Lighting **272**
    Street Lighting **264**
    Wasted Light **263**
Light Year **65**

M

Macular Degeneration **121**
Magnitude **70**
Messier (M) Objects **83**
Meteorites **85**
Meteors **85**
    Meteor Showers
        Leonids **27**
Moon
    Dark of the Moon **84**
    Lunar Occultations **86**
    Total Lunar Eclipses **5, 34**

N

Nebulas **16**
NGC & IC Objects **84**
North Celestial Pole (NCP) **75**

O

Observatories **227**
    Extremely Large Telescope **250**
    Keck Observatory **52**
    Kitt Peak Observatory **245**
    Mauna Kea Observatory **13, 52, 237**
    McDonald Observatory **245**
    Mt Graham Observatory **244**
    Mt Hopkins Observatory **244**
    Mt Palomar Observatory **53, 247**
    Mt Wilson Observatory **247**
    Observatorio Del Rogue De Los Muchachos **230**

Radio Telescopes **257**
    Aperture Spherical Radio Telescope (FAST) **259**
    APEX Millimeter and Submillimeter Telescope **258**
    Arecibo **257**
    Australia Telescope Compact Array **257**
    ESO's ALMA (Atacama Large Millimeter-submillimeter Array) **258**
    Giant Metrewave Radio Telescope **259**
    Lovell **258**
    Parkes **257**
    SKA (Square Kilometre Array) **259**
    Very Large Array (VLA) **258**
Siding Spring / the Anglo-Australian Observatory (AAO) **228**
South African Astronomical Observatory **242**
Space Telescopes **254**
    Herschel **254**
    Hubble (HST) **2, 254**
    James Webb (JWST) **255**
    Kepler **255**
    Spitzer **254**
    StarShade Planet-finding Telescope **256**
    WFIRST (Wide Field Infrared Survey Telescope) **256**
The European Southern Observatory (ESO) **232**
The Giant Magellan Telescope (GMT) **249**
Observatory Designs **201**
    Astro Haven Observatories **226**
    Commercial Double Observatories **216**
    Commercial Split Roof Observatories **216**
    Dual Roll-Off Roof Observatories **207**
    Interior Design **211**
    Roll-Away Shed Observatories **207**
    Roll-Off Roof Observatories **204**
    Skyshed Pod Observatory **203**
    Split Shed Roll-Apart Observatories **218**
    Traditional Dome Observatories **201**
Observing **97, 104, 110, 123, 124, 125, 132**
    Comfort **104**
    Effects of Body Heat **107**
    Eyesight **111**
        Averted Vision **117**
        Floaters **120**
        Iris **121**
        Night Blindness **116**
        Vision Tests **111, 112, 118**
            Dark Adaption **113, 114**

Index

Good Seeing **97**, **176**
Observing from Mountains **99**, **102**
Poor Seeing **97**
Sky Brightness **102**
Stray Light **116**
Wind Protection **104**
Occultations **86**

P

Perihelion **83**
Phases **87**
Planetariums **8**

R

Refraction **87**
Right Ascension **77**

S

Scintillation **89**
Sir Isaac Newton **2**
Skyscout **67**
Solar Systems **15**
Solstices **81**
South Celestial Pole (SCP) **76**
Spectrographs **72**
Speed of Light **65**
Star Atlases **68**
    Touring The Universe Through Binoculars Atlas' (TUBA) **126**
Star Party **10**, **11**

Stars **15**
Sun
    Total Solar Eclipses **4**

T

Telescopes **140**
    Aperture **143**
    Collimation **144**
    Eyepieces **147**
    Filters **148**
    Focal Ratio & Focal Length **145**
    Magnification **146**
    Mounts **149**, **152**, **206**
    Newtonian Reflecting Telescopes **140**
    Observing Chair **154**
    Occulting Bar **148**
    Refracting Telescopes **140**
    Schmidt-Cassegrain Reflecting Telescopes **140**
Terminator **85**
Torchlight **116**
Transits **87**
Twilight **89**

U

Universe
    Filamentary Structure **17**

Z

Zenith **81**
Zodiac **81**

Printed in Great Britain
by Amazon